Lecture Notes in Mathematics

Edited by A. Dold and B. Eckmann

951

Advances in Non-Commutative Ring Theory

Proceedings of the Twelfth George H. Hudson Symposium
Held at Plattsburgh, USA, April 23–25, 1981

Edited by P.J. Fleury

Springer-Verlag
Berlin Heidelberg New York 1982

Editor
Patrick J. Fleury
SUNY – Plattsburgh, Faculty of Arts and Science, Department of Mathematics
Plattsburgh, New York 12091, USA

AMS Subject Classifications (1980): 16–02, 16–06, 16 A 04, 16 A 08, 16 A 14, 16 A 33, 16 A 34, 16 A 38, 16 A 45, 16 A 52, 16 A 60

ISBN 3-540-11597-8 Springer-Verlag Berlin Heidelberg New York
ISBN 0-387-11597-8 Springer-Verlag New York Heidelberg Berlin

Printing and binding: Beltz Offsetdruck, Hemsbach/Bergstr.
2146/3140-543210

Preface

These papers are the proceedings of the Twelfth George H. Hudson Symposium: Advances in Non-Commutative Ring Theory which was held by the Department of Mathematics of the State University College of Arts and Science at Plattsburgh, New York, and which took place on April 23-25, 1981.

The conference consisted of talks by five invited speakers and thirteen other speakers who contributed papers, and in this volume we have collected papers by two of the invited speakers and seven of the contributors. While not all of the papers given at the Symposium appear in this volume, some of the contributors have taken the opportunity to elaborate on their contributions.

At this time, the organizers of the Symposium would like to express their thanks to the following:

The National Science Foundation and, especially, Dr. Alvin Thaler for support under NSF Grant MC580-1655.

Dean Charles O. Warren and Mr. Robert G. Moll of the Dean's Office for expert administrative support.

The Mathematics Department at PSUC and its chairman, Dr. Robert Hofer, for moral support and a great deal of hard work.

Ms. Carol Burnam, secretary par excellence, without whom the entire project would have fallen into chaos many times over.

Finally, to Dr. Paul Roman, Dean of Graduate Studies and Research at PSUC who supplied excellent advice, unstinting support, vast amounts of time, and a great deal of encouragement, we can only give a very inadequate "Thank you."

P. Fleury

Plattsburgh, N.Y.

List of Participants

Name	Institution
Maurice Auslander	Brandeis University
John Beachy	Northern Illinois University
Gary Birkenmeier	Southeast Missouri State University
William Blair	Northern Illinois University
Hans Brungs	University of Alberta
Lindsay Childs	State University of New York, Albany
Miriam Cohen	Ben Gurion University of the Negev
Paul M. Cohn	Bedford College, University of London
Robert Damiano	George Mason University
John Dauns	Tulane University
Richard Davis	Manhattan College
Warren Dicks	Syracuse University
Carl Droms	Syracuse University
Carl Faith	Rutgers University
Syed M. Fakhruddin	University of Petroleum and Minerals
Theodore Faticoni	University of Connecticut
Jose Gomez	I.B.M.
Edward Green	Virginia Polytechnic Institute and State University
John Hanna	University College, Dublin
Allan Heinicke	University of Western Ontario
Yehiel Ilamed	Soreq Nuclear Research Centre
Marsha Finkel Jones	University of North Florida
Jeanne Kerr	University of Chicago
Jacques Lewin	Syracuse University
Peter Malcolmson	Wayne State University
Wallace Martindale	University of Massachusetts
Gordon Mason	University of New Brunswick
Robert Raphael	Concordia University
Idun Reiten	University of Trondheim
Richard Resco	University of Oklahoma
J. Chris Robson	Leeds University
Jerry Rosen	University of Massachusetts
Mary Rosen	University of Massachusetts
William Schelter	University of Texas
Jan Van Geel	University of Antwerp
John Zeleznikow	Michigan State University

Participants from State University College of Arts and Science, Plattsburgh, New York

Joseph Bodenrader
Lonnie Fairchild
William Hartnett
Romuald Lesage
Kyu Namkoong
John Riley
Paul Roman
Ranjan Roy
Wei-Lung Ting
Donald C. West

TABLE OF CONTENTS

Invited Speakers

Contributing Speakers

TORSION MODULES AND THE FACTORIZATION OF MATRICES

P. M. Cohn
Department of Mathematics, Bedford College,
Regent's Park, London NW1 4NS.

1. For firs (and even semifirs) there is a fairly complete factorization theory for elements and more generally for square matrices. In terms of modules this leads to the category of torsion modules, and two questions arise naturally at this point:

1. Do these or similar results hold for more general rings?

2. What can be said about the factorization of rectangular matrices?

Below is a progress report. It turns out that torsion modules can be defined over very general rings (weakly finite rings), but as soon as we ask for more precise information we are hampered by the lack of a good factorization theory, which so far is missing even for the semifirs' nearest neighbours, the Sylvester domains. The basic results on the factorization of rectangular matrices are stated here, but some shortcomings will be pointed out, which will need to be overcome in a definitive treatment.

2. If R is a principal ideal domain, it is well known that any submodule of R^n has the form R^m with $m \leqslant n$. So any finitely generated R-module M has a resolution

$$(1) \qquad 0 \to R^m \to R^n \to M \to 0,$$

and $n - m$ is an invariant, the <u>characteristic</u> of M, written $\chi(M)$. By what has been said, $\chi(M) \geqslant 0$ always; the modules M with $\chi(M) = 0$ are just the <u>torsion modules</u>.

An obvious generalization is to take rings in which each submodule of a free module is free, of unique rank. These are just the firs (= free ideal rings, cf. [2], Ch. 1), e.g. the free algebra $k\langle X\rangle$ on a set X over a field k. But there is an important difference, in that we can now have $\chi(M) < 0$; e.g. when $R = k\langle X\rangle$,

$M = R/(Rx + Ry)$ has $\chi(M) = -1$. To find an analog to the PID case we need the notion of a <u>positive module</u>. This is a module M such that $\chi(M') \geq 0$ for all submodules M' of M. If M is positive and $\chi(M) = 0$, we call M a <u>torsion module</u>. As the presentation (1) shows, M is then defined by a square matrix A, and the positivity of M means that A is <u>full</u>, i.e. we cannot write $A = PQ$, where P has fewer columns than A.

For completeness we define a <u>negative module</u> as a module M such that $\chi(M'') \leq 0$ for all quotients M" of M. It is not hard to see that there is a duality (= anti-equivalence) between the category of all negative left R-modules and the category of all positive right R-modules such that $\mathrm{Hom}_R(M,R) = 0$ (the <u>bound</u> modules), for any fir, or more generally, any semifir (cf. [3]). A module is said to be <u>prime</u> if M is either positive and $\chi(M') > 0$ for any non-zero submodule M', or negative and $\chi(M'') < 0$ for any non-zero quotient M". As an example of a prime module of characteristic 1 we can take the semifir R itself. Now we have

Proposition 1 (cf. [4]). <u>If R is a semifir and M, N are prime R-modules of characteristic 1, then any non-zero homomorphism</u> $f: M \to N$ <u>is injective</u>.

Proof. We have the exact sequence

$$0 \to \ker f \to M \xrightarrow{f} N \to \mathrm{coker}\, f \to 0.$$

If $\ker f \neq 0$, then $\chi(\ker f) > 0$, so $\chi(\mathrm{coker}\, f) > 0$, $\chi(\mathrm{im}\, f) = 1 - \chi(\mathrm{coker}\, f) \leq 0$, hence $\mathrm{im}\, f = 0$.

From the Proposition we easily obtain the

Corollary. <u>If M is a prime module of characteristic 1 over a semifir, then</u> $\mathrm{End}_R(M)$ <u>is an integral domain.</u>

Let me outline, following G.M. Bergman [1], how Prop. 1 can be used to show the existence of a field of fractions for a fir. Consider all the prime left R-modules of characteristic 1 extending R. They form a category which is a partial ordering: two homomorphisms $M \to N$ agreeing on R must be equal, by Prop. 1. The category 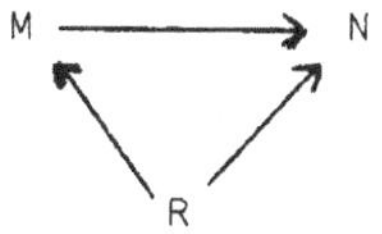

is directed since we can form pushouts (it is at this point that one needs firs rather than semifirs). Let L be the direct limit, then $End_R(L)$ contains R (via right multiplications), and it is a skew field, because the set of all endomorphisms is transitive on non-zero points.

3. We now examine what assumptions on the ring are really needed in the preceding development. To begin with, let R be any ring, ${}_RP$ the class of all finitely generated projective left R-modules and $K_0(R)$ the projective module group, with generators [P], for $P \in {}_RP$, and defining relations $[P \oplus Q] = [P] + [Q]$. As is well known (and easily seen), each element of $K_0(R)$ has the form $[P] - [Q]$ and $[P] - [Q] = [P'] - [Q']$ if and only if

$$P \oplus Q' \oplus T \cong P' \oplus Q \oplus T \text{ for some } T \in {}_RP. \tag{2}$$

Here we may of course replace T by R^n.

We define a partial preorder, the <u>natural preorder</u> on $K_0(R)$ by putting

$$[P] - [Q] > 0 \text{ whenever } [P] = [Q] + [S] \text{ for some } S \in {}_RP. \tag{3}$$

Our first concern is to know when this is a partial order:

Proposition 2. <u>The natural preorder on</u> $K_0(R)$ <u>is a partial order if and only if</u>

$$S \oplus T \oplus R^n \cong R^n \Rightarrow S \oplus R^m \cong R^m. \tag{4}$$

For we have a partial order if and only if $[P] \geq [Q] \geq [P]$ implies $[P] = [Q]$, i.e. $[S] \leq 0 \Rightarrow [S] = 0$, and this is just (4).

We recall that a ring R is said to be <u>weakly finite</u> if for any square matrices of the same size, $AB = I \Rightarrow BA = I$, or equivalently, $P \oplus R^n = R^n \Rightarrow P = 0$ (other names: R_n for all n, is v. Neumann finite, directly finite, inverse symmetric). It is clear that in a weakly finite ring (4) holds, so we have

Theorem 1. _In any weakly finite ring_ R _the natural preorder on_ $K_0(R)$ _is a partial order and_ $[P] = 0 \Rightarrow P = 0$.

To define torsion modules we have to limit ourselves to modules with a finite resolution. Let us call a module M finitely resolvable if it has a finite resolution by finitely generated projective R-modules:

$$0 \to P_n \to \cdots \to P_1 \to P_0 \to M \to 0, \qquad (P_i \in {}_R P). \tag{5}$$

Write ${}_R\overline{P}$ for the class of all such M. Given two finite resolutions of M, say (5) and

$$0 \to Q_n \to \cdots \to Q_1 \to Q_0 \to M \to 0, \tag{6}$$

(without loss of generality both are of the same length, we have by the extended Schanuel-lemma (cf. [6], p. 137)

$$P_0 \oplus Q_1 \oplus P_2 \oplus \cdots \cong Q_0 \oplus P_1 \oplus Q_2 \oplus \cdots$$

Hence the alternating sums for the sequences (5) and (6) define the same element of $K_0(R)$ and we can define the characteristic of M by the formula

$$\chi(M) = \Sigma(-1)^i[P_i]. \tag{7}$$

Starting from any resolution (5) of M, we can modify $P_1, \cdots, P_{n-1}$ so that they become free of finite rank. If in this case the last module P_n is also free, M is said to have a _finite free resolution._ Clearly when this is so, we have $\chi(M) = n[R]$ for some $n \in Z$ (this holds more generally whenever the last term P_n in the above resolution is stably free).

It is easily seen (and well known) that $\chi(M)$ is additive on short exact sequences: Given a short exact sequence

$$0 \to M' \to M \to M'' \to 0,$$

if two of M, M', M" are in ${}_R\overline{P}$, then so is the third, and we have $\chi(M) = \chi(M') + \chi(M'')$.

We can now define for any $M \in {}_R\overline{P}$:

1. M is _positive_ if $\chi(M') \geq 0$ for all submodules M' of M in ${}_R\overline{P}$.
2. M is _negative_ if $\chi(M'') \leq 0$ for all quotients M" of M in ${}_R\overline{P}$.
3. M is a _torsion module_ if it is both positive and negative.
4. M is _prime_ if either M is positive and $\chi(M') > 0$ for non zero submodules M' or M is negative and $\chi(M'') < 0$ for non-zero quotients M".

Now it is clear that Prop. 1 holds for any projective free ring (i.e. a ring over which every finitely generated projective module is free, of unique rank). More generally, a similar result will hold for any ring with a minimal positive projective module.

As in the case of semifirs (cf. [2], Th. 5.3.3, p. 185) one now has

Theorem 2. _For any weakly finite ring_ R _the torsion modules form an abelian category_ T _which is a full subcategory of R-Mod_.

The proof follows closely the semifir case, using the natural ordering in $K_0(R)$, and the following criterion (cf. [2], Prop. A.3, p. 321. I am obliged to C.M. Ringel for drawing my attention to an omission in the enunciation, which is rectified below).

Let A _be an abelian category and_ B _a full subcategory; then_ B _is abelian if and only if it has finite direct sums and the kernel and cokernel (taken in_ A) _of any map in_ B _lie again in_ B.

4. Over a commutative Noetherian ring every torsion module is annihilated by a non-zero divisor (Auslander-Buchsbaum theorem, cf. [6], p. 140). This is certainly no longer true in general, e.g. R/xR, where R = k<x,y>, is a torsion module whose annihilator is 0, but it may well extend to non-commutative Noetherian

domains.

When we come to look at general (weakly finite) rings, one difficulty is the paucity of prime modules. We saw that for a semifir R, R itself is prime. Below we examine another, wider, class of rings for which this is true.

We recall that for any matrix A (over any ring) the <u>inner rank</u> of A, rk A, is defined as the least r such that A = PQ, where P has r columns. Now Dicks and Sontag [5] have defined a <u>Sylvester domain</u> as a ring R such that

(8) $$A\ m \times r,\ B\ r \times n,\ AB = 0 \Rightarrow \text{rk}A + \text{rk}B \leq r.$$

The reason for the name is that (8) is a special case of Sylvester's law of nullity:

(9) $$\text{rk}\ A + \text{rk}\ B \leq r + \text{rk}\ AB,$$

for A m x r, B r x n. Conversely, we can deduce Sylvester's law (9) from (8). For if AB in (8) has inner rank s, say AB = CD, where D has s rows, then $(A,C)\begin{pmatrix}-B\\ D\end{pmatrix} = 0$, hence rk A + rk B $\leq$ r + s, i.e. (9). Any Sylvester domain has a universal field of fractions inverting all full matrices, in fact this property can be used to characterize them; thus Sylvester domains include semifirs. Further, any Sylvester domain is projective free, of weak global dimension at most 2. For an Ore domain the converse holds: any projective free Ore domain of weak global dimension at most 2 is a Sylvester domain. E.g. k[x,y] is a Sylvester domain, but not k[x,y,z].

Proposition 3. <u>For any coherent Sylvester domain</u> R, R <u>is a prime module</u>.

Proof. We must show that for every finitely presented non-zero left ideal a of R, X(a) > 0.

Let a be generated by $c_1,\dots,c_n$ and take a resolution

(10) $$0 \to F \xrightarrow{\alpha} R^n \to a \to 0.$$

We note that w.dim(R/a) $\leqslant 2$, hence w.dim(a) $\leqslant 1$, so the first term F in (10) is flat; by coherence it is finitely generated, hence finitely presented, therefore projective, and so free (because R is projective free). If α has a matrix $A = a_{ij})$, then $Ac = 0$, where $c = (c_1, \cdots, c_n)^T = 0$, hence rk A + rk c $\leqslant$ n, but rk c $\geqslant$ 1, so rk A $\leqslant$ n - 1. Thus A = PQ, where P is m x p, Q is p x n and $p \leqslant n - 1$. It follows that PQc = 0 and rk P = p = rk Q, and rk P + rk Qc $\leqslant$ p, hence Qc = 0. Moreover, Qx = 0 implies Ax = 0, hence we have a presentation of a by Q instead of A and $X(\mathbf{a}) = n - p \geqslant 1$.

5. It looks at first sight as if much of the theory of semifirs carries over to Sylvester domains, but we run into difficulties as soon as we consider the factorization (of elements or matrices) over Sylvester domains. To make a beginning let us see how the factorization theory of semifirs treated in Ch. 5 of [2] extends to rectangular matrices. In Ch. 5 of [2] there is a factorization theory for square matrices over semifirs, but nothing beyond a few remarks (on p. 202f.) about rectangular matrices.

Let R be any ring, then any matrix $A \in {}^mR^n$ defines a module M:

$$(11) \qquad R^m \to R^n \to M \to 0,$$

where the map α has matrix A, and A is determined by M if and only if $xA = 0 \Rightarrow x = 0$, i.e. A is a right non-zerodivisor. We remark that A is a left non-zerodivisor if and only if $M^* = \mathrm{Hom}_R(M,R) = 0$, i.e. M is a bound module. We also note that M = 0 if and only if A has a left inverse.

When R is a semifir, every finitely presented module M is defined by a right non-zerodivisor matrix A, i.e. α in (11) is then injective. In that case X(M) = n - m; we shall also call n - m the characteristic of the matrix A: char A = n - m.

Two matrices A, A' define isomorphic modules if and only if they are <u>stably associated</u>, i.e. $\begin{pmatrix} A & 0 \\ 0 & 1 \end{pmatrix} = U \begin{pmatrix} A' & 0 \\ 0 & 1 \end{pmatrix} V$ for invertible matrices U, V (where the unit matrices need not be of the same size.) Conversely, every matrix A which is a right non-zerodivisor defines a left module M, and a matrix product $C = AB$ corresponds to a short exact sequence

$$0 \to M' \to M \to M'' \to 0,$$

where A, B, C define M', M", M respectively. More generally, if we consider all factorizations of a full matrix C, there is a correspondence between the left and right factors, the factorial duality (cf. [2], p. 119), which means for example, that an integral domain which satisfies the ascending chain condition on principal right ideals (right ACC_1 for short), also satisfies the descending chain condition on principal left ideals containing a given non-zero left ideal. We recall that a fir satisfies right ACC_n, i.e. ACC on n-generator right ideals, for any n ([2], p. 49) by an atom in a ring we mean a non-unit which cannot be written as a product of two non-units. Now the factorization theorem for firs may be stated as follows:

Theorem (cf. [2], p. 201). *In an n × n matrix ring over a fir every full matrix can be written as a product of atoms, and given two factorizations into atoms:*

(12) $$C = A_1 \cdots A_r = B_1 \cdots B_s,$$

we have $r = s$ *and there is a permutation* $i \to i'$ *such that* A_i *is stably associated to* $B_{i'}$.

Here all the matrices are n × n over the ground ring, for some fixed n. We are interested in the generalization to the case where the A's and B's are not necessarily square and even C need not be square. For this purpose we have to examine more closely the steps by which one passes from one factorization of C in (12) to another.

We recall that a relation between matrices

(13) $$AB' = BA'$$

is called _comaximal_ if (A,B) has a right inverse and $\begin{pmatrix} B' \\ A' \end{pmatrix}$ a left inverse. Let A be r × m, B r × n, A' n × s and B' m × s, then by the law of nullity in semifirs,

$$r + s \leq m + n. \tag{14}$$

If equality holds in (14), so that char A = char A' = m - r, char B = char B' = n - r, we call (13) a _proper_ comaximal relation. Thus for any comaximal relation C = AB' = BA' over a semifir we have

$$\text{char } C \leq \text{char } A + \text{char } B,$$

with equality if and only if the relation is proper. Now one has

Lemma 1 ([4], Prop. 2.2). _Two matrices_ A, A' _over a weakly finite ring_ R _are stably associated if and only if there is a proper comaximal relation (13) for_ A, A'.

If in some factorization a product AB' is replaced by BA', where AB' = BA' is a (proper) comaximal relation, we shall call the change a _(proper) comaximal transposition_. This extends the usage in [2], p. 134.

Now we have

Theorem 3 (Refinement theorem). _Let_ R _be a semifir and_ C ε $^mR^n$, _then any two factorizations of_ C _have refinements which can be obtained from each other by comaximal transpositions_.

The proof, which is quite straightforward, is analogous to the corresponding refinement theorem for factorizations into square matrices. However, this theorem does not seem to be in the best possible form in that we cannot always choose the comaximal transpositions to be proper. This happens (roughly speaking) when C is too narrow in shape, i.e. of large positive or negative characteristic. If we translate this into module language we find that comaximal relations correspond to sums and intersections, but when the relation is improper, the intersection contains a free summand.

In order to state a factorization theorem we need to find an analog of atoms for rectangular matrices. Let us call a matrix factorization $C = AB$ _proper_ if A has no right inverse and B no left inverse. If C is neither a unit nor a zero-divisor and has no proper factorizations, then we call it _unfactorable_.

It is easily seen that a matrix C has a proper factorization if and only if the module M defined by it has a proper non-zero bound submodule. This leads to the following description of the modules defined by unfactorable matrices:

Proposition 4. _Let_ R _be a semifir, then a finitely presented R-module_ M _has an unfactorable matrix if and only if every proper finitely generated submodule of_ M _is free_.

Proof. Suppose that M has a proper bound submodule $M' \neq 0$, then M' is clearly not free. Conversely, if M' is a non-free proper submodule of M, either M' is bound and we have a proper factorization, or $M'^* \neq 0$, so there is a non-zero homomorphism $F: M' \to R$. Its image (finitely generated as image of M') is free, as submodule of R, and hence splits off M': $M' = F \oplus M_1'$. By induction on the number of generators M_1' has a bound non-zero submodule and the result follows.

Sometimes a module M is called _almost free_ if M is not free, but every proper finitely generated submodule is free. Thus unfactorable matrices correspond to almost free modules. However, we shall not pursue the module aspect here further.

To prove the factorization theorem we isolate the essential step in the follow g basic lemma:

Lemma 2. _Let_ R _be a semifir and_ C _any matrix over_ R. _Given_

$$C = AB' = BA',$$

where A _is unfactorable and_ BA' _is a proper factorization, either there exists a matrix_ U _such that_ $B = AU$, $B' = UA'$, _or there is a comaximal relation_ $AB_1 = BA_1$ _such that_ $A' = A_1Q$, $B' = B_1Q$, _for some matrix_ Q.

The proof is quite similar to the corresponding result for elements

([2], p. 124f.). With the help of this lemma we obtain

Theorem 4 (Factorization theorem). Let R be a fir, then every matrix C over R which is a non-zerodivisor has a proper factorization into unfactorables, and given any two such factorizations of C, we can pass from one to the other by a series of comaximal transpositions.

The existence of factorizations was proved in [2], Th. 5.6.5, p. 202, and the uniqueness follows by repeated application of Lemma 2.

REFERENCES

[1] G.M. Bergman, Dependence relations and rank functions on free modules, to appear.

[2] P.M. Cohn, Free rings and their relations, LMS Monographs No. 2, Academic Press (London, New York 1971).

[3] P.M. Cohn, Full modules over semifirs, Publ. Math. Debrecen, 24 (1977), 305-310.

[4] P.M. Cohn, The universal field of fractions of a semifir I. Numerators and denominators, Proc. London Math Soc. (in press).

[5] W. Dicks and E.D. Sontag, Sylvester domains, J. Pure Applied Algebra 13 (1978), 143-175.

[6] I. Kaplansky, Commutative rings, Allyn and Bacon (Boston (1970).

SUBRINGS OF SELF-INJECTIVE AND FPF RINGS

For Molly Sullivan Wood

Carl Faith

Abstract

We say that a ring K is <u>(right) split by a subring</u> A provided that A is an (right) A-module direct summand of K. Then K is said to be <u>a split extension of</u> A. By a theorem of Azumaya [1], a necessary and sufficient condition for this to happen is that K generates the category mod-A of all right A-modules. A classical example of this occurs when $A = K^G$ is a Galois subring corresponding to a finite group of invertible order $|G|$. In order that A be a right self-injective subring of K it is necessary that A split in K, and the latter condition is sufficient for a right self-injective left A-flat extension K of A (Theorem 1).

We also study when the (F)PF property is inherited by a subring A: K is right (F)PF if each (finitely generated) faithful right K module generates mod-K. Any quasi-frobenius (QF) ring is right and left PF; any commutative Prufer domain, and any commutative self-injective ring is FPF [4,5].

The main theorem on FPF rings states that A inherits the right (F)PF hypothesis on K when K is left faithfully flat right projective generator over A. Now another theorem of Azumaya [1] states that if A is commutative, then any finitely generated faithful projective A-module generates mod-A, hence a corollary is that K FPF => A FPF whenever K is finitely generated projective over a commutative subring A.

We apply the foregoing results to a subring A of a right self-injective ring K in the case that A is right non-singular. Then, assuming that $_AK$ is flat, by the structure theory of nonsingular rings K (being injective over A on the right) contains a unique injective hull of A which is canonically the maximal quotient ring $Q = Q_{max}(A)$, and, moreover, then Q splits in K (Theorem 4.) This holds in particular if A is a von Neumann regular ring (Corollary 5). Furthermore, if $A = K^G$ is a Galois subring, then $A = Q^r_{max}(A)$ is right self-injective (Theorem 6 and Corollary 7).

As a final application we derive a theorem of Armendariz-Steinberg [19] stating that if K is a right self-injective regular ring then the center of K is self-injective (Theorem 10).

PROOFS OF THEOREMS

1. THEOREM. If K is left flat over A, and right self-injective, then A is right self-injective iff A splits in K.

Proof. Since the left adjoint $\otimes_A K$ of the inclusion functor I: mod-K $\rightsquigarrow$ mod-A is exact, then (e.g., by Theorem 6.28 of [13a]) I preserves injectives, i.e., K_A is injective. If K generates mod-A, this implies that A_A is injective. Conversely, A_A injective implies that K/A is split.

EXAMPLE: K is commutative, self-injective and subring A splits in K, yet A is not self-injective:

The example is the split-null extension K = (A,E) of a balanced injective module E. For example, E can be Z_{p^∞}, and A = End Z_{p^∞} the right of p-adic numbers. (See [11] for details of this and the following.) Then K = (A,0) $\oplus$ (0,E), so A $\approx$ (A,0) splits in K, but A is a domain not a field, so is not self-injective.

A module M over a ring R is <u>right co-faithful</u> provided that there is an integer $n < \infty$ and an embedding $R \hookrightarrow M^n$. Clearly any generator is co-faithful; any finitely generated module M over a commutative ring R is co-faithful, since if $x_1, \cdots, x_n$ generate M, the mapping $R \to M^n$ sending $r \in R$ onto $(x_1 r, \cdots, x_n r) \in M^n$ is an embedding.

A ring R if right (F)CF provided that every (finitely generated) faithful right R-module is co-faithful.

By definition, any right (F)PF ring R is right (F)CF. Moreover, as just shown, any commutative ring R is FCF. It can be shown that any right semi-artinian ring is right CF, and hence any right artinian ring is right CF. (See [1] for example, or [13b], Chapter 19).

We say that K is a (right) projective extension of A if K is a projective object of mod-A. Then if K/A splits, we say that K/A is <u>split-projective</u>.

2. THEOREM. If K/A is split-projective and if K is right (F)PF, then A is right (F)PF iff A is right (F)CF.

Proof. Any right (F)PF ring is right (F)CF. Conversely. Suppose that K is right (F)PF, and let M be any (f.g.) faithful, hence co-faithful right A-module. Since $A \hookrightarrow M^n$, then $K \hookrightarrow M^n \otimes_A K$, so $M \otimes_A K$ is faithful over K. By (F)PF, $M \otimes_A K$ generates mod-K, and hence generates mod-A. Write $K \oplus X = A^{(a)}$ for a cardinal number a. Then

$$M \otimes_A K \oplus X \otimes_A K = M \otimes_A A^{(a)} \approx M^{(a)}$$

and since $M \otimes_A K$ generates mod-A, so does $M^{(a)}$, hence M. This proves that A is (F)PF.

2A. Corollary. If K is split-projective over a commutative subring A, then A is FPF if K is. Moreover, A is self-injective if K is.

Proof. Any commutative ring is FCF, so Theorem 2 suffices for the FPF part. Moreover, Theorem 1 suffices for the injective part.

2B. Corollary. If K is finitely generated projective over a commutative subring A, then K right FPF implies that A is FPF.

Proof. By a theorem of Azumaya [1] any finitely generated projective faithful module over a commutative ring A generates mod-A, so the last corollary applies.

R is right CFPF if every factor ring R/I is FPF. Commutative (C)FPF rings have been classified [5].

2C. Corollary. If K is a commutative CFPF ring, and if K is a projective generator over a subring A, then A is CFPF.

Proof. Let I be any proper ideal of A. Since $K \approx A \oplus X$ in mod-A, then $KI \approx I \oplus XI$, hence $K/KI \approx A/I \oplus X/XI$ in mod-A. This proves that K/IK generates mod-A/I. Similarly, one shows that K/IK is projective over A/I. Since K is CFPF, then K/IK is FPF, and hence by Corollary 2A, so is A/I.

3. THEOREM. If K is a split-flat extension of A, and if K is QF, then so A. Moreover, K is then split-projective-injective over A (both sides).

Proof. By Theorem 1, A is right self-injective. Since K is right Noetherian, K satisfies acc $\perp$, namely the acc on annihilator right ideals, and hence, so does any subring. In particular, A satisfies acc $\perp$, hence A is QF by a theorem of [15]. Now, every injective is projective over a QF ring A, so K is projective in mod-A by the proof of Theorem 1. Now over a left perfect ring, every flat module is projective, so K is left projective over A. Finally, since ${}_AA$ is injective, K is left split over A.

APPLICATIONS TO GALOIS THEORY

We begin with a theorem on a nonsingular subring.

4. THEOREM. Let K_K be injective, and let A be a nonsingular subring such that ${}_AK$ is flat. Then K contains the maximal quotient ring $Q = Q^r_{max}(A)$. Moreover, K is a split extension of Q.

Proof. K_A is injective, so K contains an injective hull E = E(A) of A in mod-A. But then E is a ring ≈ Q by well known theorems of Johnson-Utumi. Since Q is right self-injective, then Q splits in K.

5. Corollary. If K_K is injective, and if A is a regular subring, then K contains $Q = Q^r_{max}(A)$.

Proof. For then ${}_AK$ is flat.

6. THEOREM. If K_K is injective, ${}_AK$ is flat, and if $A = K^G$ is a right non-singular Galois subring of K, then $A = Q^r_{max}(A)$ is right self-injective and regular.

Proof. Any injective A-module, e.g. K, contains a unique copy of E = E(A). (See for example, my Lectures, [17], p. 62, Theorem 8.) Now g maps E onto g(E), and since $g(xa) = g(x)a \ \forall a \in A$, then $g(E)$ is an injective hull of A, so $g(E) = E$ by uniqueness. This proves that g induces an automorphism $\bar{g}$ of $Q = Q^r_{max}(R)$. However, since $\bar{g}/A = 1_A$, then $\bar{g} = 1_Q$. (This follows since for each $x \in Q$, there exists an essential right ideal I of A so that $xI \subseteq A$. Then clearly $(g(x)-x)I=0$ hence $g(x)-x = 0$, proving that $\bar{g} = 1_Q$.) This shows that $K^G = A = Q$.

7. Corollary. If K_K is injective, and if A is a von Neumann regular Galois subring, then A is right self-injective. Moreover, any maximal commutative regular subring B is self-injective.

Proof. Over a regular ring, every left A-module is flat, hence by the last corollary, $A = Q^r_{max}(A)$ is right self-injective. Corollary 5 shows that $B = Q_{max}(B)$, hence B is self-injective.

ON THE CENTER OF THE REGULAR RING

If R is regular, then its center C is regular. If R is also (right) self-injective, is C? The answer is "yes."

8. THEOREM (Armendariz-Steinberg [19]. If K is a right self-injective regular ring, then the center C of K is self-injective generated by units.

Proof. By a result of Henriksen [20], the ring K_2 of all 2×2 matrices over K is generated by units for any ring K, and K_2 has center isomorphic to C^2. Since K_2 is Morita equivalent to K, then K_2 is self-injective when K is, so to prove the theorem it suffices to assume that K is generated by units. Then, K is Galois over $C = K^G$, and since C is regular, then C is self-injective by Corollary 7.[3]

[18] contains an example $K = F_3$ of 3×3 matrices over a field $F \approx GF(2)$ with a group G of inner automorphisms (≈ Klein fours group), with $A = K^G$ a local ring not self-injective.

[18] also contains many sufficient conditions for the implication K_K injective => $A = K^G$ injective in a regular ring K holding for $|G|<\infty$: Among these are: (1) $|G|$ is a unit of K; (2) K is reduced (i.e. without nilpotent

elements $\neq 0$); $A \subseteq C$.

Remarks. 1. Note that $|G|$ a unit implies that A splits in K, so by Theorem 1 the proof devolves to showing the ${}_AK$ is flat, and for this it suffices to prove that A is regular. (${}_AK$ is actually finitely generated projective.)

2. (2) => shows that A is nonsingular, so again it devolves to showing that ${}_AK$ is flat by Theorem 6. To wit: C is self-injective, and G induces a group of automorphisms G' in C, with the same fixring as G, so it suffices to prove (3) assuming that K = C. (3) is closely related to Theorem 8.

AFTERWORD ON GALOIS THEORY WITHOUT GALOIS GROUPS

This paper began as a sequel to [6] to study Galois subrings of self-injective and/or FPF rings. Many of the theorems in this paper were first obtained for a Galois subring A of commutative ring K corresponding to a finite group G of automorphisms satisfying various properties, e.g. G independent in the sense of [6], and/or $N = |G|$ a regular element (or unit) of K, and/or K finitely generated projective over A.

The first theorems that I proved that eliminated the need for G were the ones for K a free A-module of finite rank over A. Then, in answering my query, S. Endo pointed out how to extend this to K finitely generated projective over A. I am indebted to Professor Endo for inspiring the generality of this paper, and for providing many interesting extensions of my theorems on pre-Galois extensions.

FOOTNOTES

[1]After this was written and circulated in preprint form, I received several interesting communications on Theorem 8. P. Menal and J. Moncasi pointed out Henriksen's theorem, obviating the hypothesis that K be generated by units; E. Armendariz cited the prior result [19]; and K.R. Goodearl proved the result without recourse to K_2, [19], or [20].

[2]Menal and Moncasi also pointed out that in K_2 we have

$$\begin{pmatrix} a & b \\ c & d \end{pmatrix} = \begin{pmatrix} a & 1 \\ -1 & 0 \end{pmatrix} + \begin{pmatrix} 0 & -1 \\ 1 & d \end{pmatrix} + \begin{pmatrix} 1 & 0 \\ c & 1 \end{pmatrix} + \begin{pmatrix} 1 & b \\ 0 & -1 \end{pmatrix}$$

so every element is a sum of 4 units. In K_n, in general one can get by with three units ([20]), $n > 1$.

[3]I conjecture that Theorem 8 cannot be extended to a general self-injective ring. This is based on a study of when a split null extension $R = (B,E)$ of a bimodule E over a ring B is self-injective. By [11] this happens iff E_B is injective, and $B = \text{End } E_B$; and then the center C of R, namely the split null extension (D,F) would have to be injective, where D = center B, and F is the "center' of the module E, i.e. F = the set of all x in E so that $bx = xb$ for all b in B. Thus, F would have to be injective over D, and $D = \text{End } F_D$.

References

1. Azumaya, G., Completely faithful modules and self-injective rings, Nagoya Math. J. 27 (1966) 697-708.

2. Endo, S., Completely faithful modules and quasi-Frobenius rings, J. Math. Soc. Japan 19 (1967) 437-456.

3. Tachikawa, H., A generalization of quasi-Frobenius rings, Proc. Amer. Math. Soc. 20 (1969) 471-476.

4. Faith, C., Injective quotient rings of commutative rings I, in Module Theory, Lecture Notes in Math. (Springer) vol. 700 (1979).

5. ________, Injective quotient rings of commutative rings II, in Injective Modules and Injective Quotient Rings, Lecture Notes in Pure and Applied Math. Vol. 72 (1982). (Dekker)

6. ________, On the Galois Theory of commutative rings, I: Dedekind's theorem on the independence of automorphisms revisited, preprint presented at the Yale symposium in honor of Nathan Jacobson, June, 1981. Contemporary Math. (to appear).

7. Auslander, M., and Goldman, O., The Brauer group of a commutative ring, Trans. Amer. Math. Soc. 97 (1960) 367-409.

8. Chase, S.U., Harrison, D.K., and Rosenberg, A., Galois Theory and Galois cohomology of Commutative rings, Memoirs of the Amer. Math. Soc. 52 (1965) 15-33.

9. Lambek, J., Rings and Modules, Blaisdell, 1974, Reprinted Chelsea, Waltham 1976.

10. Faith, C., Injective cogenerator rings and a theorem of Tachikawa I, II, Proc. Amer. Math. Soc. 60 (1976) 25-30; 62 (1977) 15-18.

11. ________, Self-injective rings, Proc. Amer. Math. Soc. 77 (1979) 157-164.

12. ________, Galois extensions of commutative rings, Math. J. Okayama U. 18 (1976) 113-116.

13a. ________, Algebra I: Rings, Modules and categories, Grundl. der Math. Wiss Bd. 190, Springer-Verlag, Berlin-Heidelberg-New York, Corrected Reprint, 1981.

13b. ________, Algebra II: Ring Theory, Grundl. der Math. Wiss Bd. 191, Springer, 1976.

14. Vamos, P., The decomposition of finitely generated modules and fractionally self-injective rings, J. London Math. Soc., (2), 16 (1977) 209-220.

15. Faith, C., Rings with ascending conditions on annihilators, Nagoya Math. J. (1966).

16. ________ and Walker, Direct sum representations of injective modules, J. Algebra (1967).

17. Faith, C., Lectures on Injective Modules and Quotient Rings, Springer Lecture Notes in Mathematics, vol. 49, Berlin, Heidelberg, and New York, 1967.

18. Goursaud, J.M., Osterburg, J., Pascaud, J.L., and Valette, J., Points fixes des anneaux reguliers auto-injectifs a gauche, preprint, Dept. des math., V. de Poitiers, 1981.

19. Armendariz, E.P., and Steinberg, S., Regular self-injective rings with a polynomial identity, Trans. Amer. Math. Soc. 190 (1974) 417-425.

20. Henriksen, M., Two classes of rings generated by their units, J. Alg. 31 (1974) 182-193.

EMBEDDING MODULES IN PROJECTIVES:

A REPORT ON A PROBLEM

Carl Faith

Rutgers, The State University
New Brunswick, NJ 08903

If every injective right R-module of a ring R embeds in a free right R-module, then R is a Quasi-Frobenius (QF) ring, by a theorem of Faith-Walker, and conversely. We say that R is right (F)GF provided that every (finitely generated) right R-module embeds in a free, equivalently, in a projective right R-module. Thus, by the theorem just quoted, right GF $\Longleftrightarrow$ QF. Inasmuch as QF is a symmetric condition for right-left, then right GF $\Longrightarrow$ left GF holds. By another result of Faith-Walker [67], right and left FGF is equivalent to QF, hence commutative FGF rings are QF.

In this paper we explore the truth or falsity of the implication

(H) $$\text{right FGF} \Longrightarrow \text{QF}$$

and show that it holds under a rather lengthy list of assumptions including right Artinian (3.2), the d.c.c. on right annihilators (3.3.1), finite essential right socle (3.5A), right Noetherian (3.5C), right self-injectivity of R (3.6), and every left ideal of R is a left annulet (3.7-8).

A theorem of Johns [77] purports to prove (H) assuming R is right Noetherian, but a lemma in the proof borrowed from Kurshan is false. (H) _is_ true for right Noetherian R (3.5C), and also true when R is _left_ Noetherian (3.3.3), hence when R is left Artinian (3.3.4).

(3.6) previously has been proved by Bjork [72] and Tol'skaya [70]. We also point out that the TF condition on a ring R imposed by Levy [63] implies that its right quotient ring Q is right GF.

Suppose there exists a cardinal number c such that every right module generated by < c elements embeds in a projective. If we call this condition cGF, Menal [81] asks if there exists a cardinal c such that

$$\text{right cGF} \Longrightarrow \text{QF}$$

Note that FGF is $\aleph_0$GF, and GF is cGF for every cardinal.

In her thesis [73a], Jain connected right IF rings with FGF rings, and we give a brief account of these theorems in Section 4, together with Rutter's sharpening of Jain's theorem: _Every right FGF ring is right IF_ (Theorem 4.1).

In the converse direction, a theorem of Jain states that any right IF ring embeddable in either a right Noetherian or right perfect ring is right FGF. (Of course, a regular ring is IF but not in general FGF.)

1. INTRODUCTION. A ring R is right ORE provided that there is an embedding $R \hookrightarrow Q$ of R into a ring $Q = Q(R)$ with the properties:

(1) every regular element $x \in R$ is a unit of Q; and

(2) $Q = \{ax^{-1} | a, \text{ regular } x \in R\}$.

$Q(R)$ is then called the right quotient ring of R, and any two, Q_1 and Q_2, are isomorphic by an isomorphism preserving the embeddings $R \hookrightarrow Q_1$ amd $R \hookrightarrow Q_2$. Ore proved that R is right Ore iff for all $a \in R$, and regular $x \in R$, there exist $a_1 \in R$ and regular $x_1 \in R$ such that $xa_1 = ax_1$.
[Mnemonic: $x^{-1}a_1 = a_1 x^{-1}$.] Thus, the well-known fact: any commutative ring is Ore.

(1.1) <u>Henceforth, let</u> R <u>denote a right Ore ring.</u>

A right R-module M is torsion free (t.f.) provided that for all $m \in M$, and regular $x \in R$, we have that $mx = 0 \Longrightarrow m = 0$. An element m of M is torsion provided $mx = 0$ for some regular $x \in M$. The set $t(M)$ of torsion elements of M is a submodule, called the torsion submodule of M. (Right Ore is necessary for $t(M)$ to be a submodule. Consult Levy [63, p. 134].) Note, $t(M)$ is the largest submodule K such that M/K is torsion free.

We let f.g. abbreviate "finitely generated." Thus f.g.t.f. abbreviates finitely generated torsion free. Then R is said to be:

right TF   if every f.g.t.f. right R-module embeds in a free R-module.

Extending Gentile's Theorem for Ore domains, Levy (l.c.) showed that when R is semiprime, then R is right TF iff R is left Ore. (For emphasis, we restate our convention that R is right Ore.) Moreover, Levy proved:

(1.2)   R right TF $\Longrightarrow$ Q right (TF)

and

(1.3)   Conversely when R is left Ore.

We now consider the question raised by Levy (l.c.):

(1.4)   Does R right TF $\Longrightarrow$ Q right Artinian?

Levy proved this affirmatively when R is semiprime, and moreover, that then Q is a semisimple Artinian ring.

(1.5) In this paper, we prove Levy's conjecture assuming R is left Ore and left TF, and actually show that Q is a QF (quasi-Frobenius) ring.
(Thus, in this case, every Q-module embeds in a free module. See below.)

Thus, for example, a commutative ring R is TF iff Q is QF. Moreover, we prove the implication

(1.6) R right TF $\Longrightarrow$ Q is QF

or equivalently, the implication

(1.7) Q right TF $\Longrightarrow$ Q is QF

under any of the following hypotheses (numbered after the corresponding theorem in Section 3):

(3.1) Q is left TF (e.g., if Q is commutative)

(3.2) Q is right Artinian.

(3.3.1) Q has the d.c.c. on annihilator right ideals.

(3.3.2) Q " " " " " left ideals.

(3.3.3) Q is left Noetherian.

(3.3.4) Q is left Artinian.

(3.4) Q is semilocal right Noetherian.

(3.5) Q has finite essential right socle.

(3.5C) Q is right Noetherian.

(3.6) Q is right selfinjective.

(3,7-8) Every left ideal is an annulet (= annihilator right ideal).

(3.6) is a result of Bjork [72] and Tol'skaya [70]. To the author's knowledge in other cases the problems are open.

2. NOTATION AND BACKGROUND. We say that a ring A is right (F)GF provided that every (finitely generated) right A-module embeds in a free right A-module. Since every module over Q = Q(R) is torsion free, we see that Q is right TF iff Q is right FGF. We now state Levy's question for any ring A.

(2.1) Does A right FGF $\Longrightarrow$ A right Artinian?

(It will be seen that right FGF $\Longrightarrow$ A is right Ore, so there is no increase in generality!)

We first remark that a theorem of Faith-Walker [67] answers (2.1) affirmatively under the stronger hypothesis:

(2.2) (right GF): every right A-module embeds in a free module.

Inasmuch as every module embeds in an injective, right GF is equivalent to the requirement that every injective embeds in a free module. In this terminology the Faith-Walker Theorem states:

(2.3) A is right GF $\Longleftrightarrow$ A is QF.

By the symmetrical properties of QF rings (below), we see that

(2.4) A is right GF $\Longleftrightarrow$ A is left GF.

2.5 Definition. *A ring R is QF provided that the following equivalent conditions hold:*

QF(a) *Every right, and every left, ideal is the annihilator of a finite subset of R.*

QF(b) *Every right, and every left, ideal is an annihilator (= annulet), and R is right or left Artinian or Noetherian.*

QF(c) *R is right selfinjective and right or left Artinian or Noetherian.*

QF(d) *R is right selfinjective and satisfies the a.c.c. on right annulets, or the a.c.c. on left annulets.*

(For proof, consult, e.g., Faith and Walker [67].)

Moreover, any QF ring satisfies the chain conditions (on both sides) stated in the definition, since there is a duality

$$X \longrightarrow X^{\perp} = \{a \in R \mid Xa = 0\}$$

between right and left annulets. (We see also that QF(d) holds replacing a.c.c. by d.c.c.)

Moreover:

2.6 Theorem (Faith-Walker [67]). *If every cyclic right, and every cyclic left, R-module embeds in a free R-module, then R is QF (and conversely).*

Thus, any right *and* left FGF-ring is QF, hence GF and Artinian, so *Levy's question has an affirmative answer assuming FGF on both sides--(as stated, in particular for commutative rings).*

Since right FGF implies that every right ideal I of R is an annulet (the embedding $R/I \to R^n$ sends the coset $[1+I]$ onto $(a_1,\dots,a_n) \in R^n$, and then I is the right annihilator of $\{a_1,\dots,a_n\}$ in R), we see that every regular element of R is a unit, that is,

$$(2.7) \qquad R \text{ right FGF} \Longrightarrow R = Q(R).$$

In particular, right FGF implies R is right Ore.

In addition to the main theorem (1.5), we also prove Levy's question (2.1) under various hypotheses, in particular, assuming A is right or left Noetherian. Curiously, we are unable to extend our proofs to the other side! Furthermore, in all these cases, A must be QF, and, in fact, if we assume at the outset that A is either right or left Artinian, then right FGF $\Longrightarrow$ QF (see Theorem 3.2-3).

3. PROOFS OF THEOREMS. We already have discussed the proof of the first theorem

3.1 Theorem. *Let R be a left and right Ore ring with quotient ring Q. Then, R is left and right TF iff Q is QF.* □

3.2 Theorem. *If Q is a right Artinian right FGF ring, then Q is QF.*

Proof. By 2.5, in order to prove that Q is QF, it suffices to prove that Q is right selfinjective. To do this, it <u>suffices to show that</u>: Q <u>has a finitely generated projective cogenerator</u> (Faith-Walker [67, Theorem 4.1]).

Let $V_1,\ldots,V_n$ denote an isomorphy class of simple right Q-modules, and let E_i denote the injective hull of V_i, $i = 1,\ldots,n$. Then, $C = E_1 \oplus \cdots \oplus E_n$ is the least injective cogenerator of mod-Q. Let E denote any of the E_i, and let F be any finitely generated submodule of E. Then, by the FGF assumption, F embeds in a free module, which by the finite generation of F can be assumed to be Q^m, for a smallest integer $m > 0$. However, since F is uniform (any two nonzero submodules intersect), then actually $m = 1$. Since every finitely generated submodule F of E therefore embeds in Q, and since Q has finite length, it follows that E satisfies the a.c.c. (and d.c.c.) on finitely generated submodules, and therefore E is Noetherian. Thus, E is finitely generated, hence E embeds in a free module Q^t. By injectivity of E, then E is a summand of Q^t, so E is projective; and then $C = E_1 \oplus \cdots \oplus E_n$ is the required finitely generated projective cogenerator. □

3.3 Lemma. *Any right FGF ring Q is QF under any one of the assumptions:*

3.3.1. Q *has d.c.c. on right annulets.*

3.3.2. Q *has a.c.c. on left annulets.*

3.3.3. Q *is left Noetherian.*

3.3.4. Q *is left Artinian.*

Proof. Clearly

$$(4) \Longrightarrow (3) \Longrightarrow (2) \Longleftrightarrow (1).$$

Since every right ideal is a right annulet (see *supra* 2.7), then (1) $\Longrightarrow$ Q is right Artinian, so by 3.2, we have (1) $\Longrightarrow$ (4) which completes the proof. □

It is possible to embed every cyclic right module in a free right R-module (in fact, in R itself) without R being right FGF. This curiosity is the usual example of a local ring R with just three right ideals $R \supset J \supset 0$ which is not left Noetherian. (We will not repeat the construction but, e.g., refer to Faith [73, p. 337].) Then 3.2 shows that R is not right FGF.

A ring R is semilocal if R/rad R is semisimple Artinian, where rad R denotes the Jacobson radical of R.

3.4 Theorem. *Any semilocal right (or left) Noetherian right FGF ring Q is QF.*

Proof. Let J denote the Jacobson radical of Q. Now, since J^n is an ideal,

$$^{\perp}J \subseteq {}^{\perp}(J^2) \subseteq \cdots \subseteq {}^{\perp}(J^n) \subseteq \cdots$$

is an ascending chain of (right) ideals so Q right Noetherian implies that $^{\perp}(J^n) = {}^{\perp}(J^{n+1})$ for some integer n. Since any right ideal is a right FGF ring is a right annulet, we have that

$$J^n = J^{n+1}$$

which in a right Noetherian ring Q implies $J^n = 0$ (by Nakayama's lemma). Then, Q/J semisimple and Q right Noetherian imply that Q is right Artinian, so that (3.2) applies. (The left Noetherian case is 3.3.3 [without assuming Q semilocal].) □

3.5A Theorem. *If Q has finite essential right socle, then right FGF ⟹ QF.*

Proof. The hypothesis implies that every free module on a finite set has (finite) essential socle, and then right FGF implies this of every finitely generated module, hence of *every* module. Then the nonexistence of infinitely many orthogonal idempotents (guaranteed by the hypothesis), and Bass's theorem [60], implies that Q is left perfect. If J = rad Q, then Q/J^2 has finite essential right socle (containing J/J^2) so that J/J^2 must be finitely generated as a right module, and then a theorem of Osofsky [67] implies that Q is right Artinian, so 3.2 applies. □

3.5B Johns' Lemma. If R is right Noetherian, and if every right ideal is an annihilator, then R has finite essential right socle.

Proof. By the proof of 3.4, $J = \text{rad } R$ is nilpotent, hence, by a result of Goldie, ${}^{\perp}J$ is an essential right ideal (Lemma 2 of Johns [77]), hence

$$^{\perp\perp}J \subseteq Z = \text{sing } R_R.$$

Since the right singular ideal Z in a ring R satisfying acc$\perp$ is nil, then $Z \subseteq J$, so ${}^{\perp\perp}J \subseteq J$, and ${}^{\perp\perp\perp}J \supseteq {}^{\perp}J$, and so

$$^{\perp}J \subseteq {}^{\perp\perp\perp}J \subseteq \ldots$$

is an ascending chain so that the m-th left annihilator $\ell^m(J)$ equals* the $(m+2)$-d left annihilator $\ell^{m+2}(J)$ for some odd m; consequently

$$\ell^{m+2}(J)^{\perp} = \ell^m(J)^{\perp}$$

so

$$\ell^{m+1}(J)^{\perp} = \ell^{m-1}(J)^{\perp}$$

etc., to obtain

$$^{\perp}J = \ell(J) = J^{\perp} .$$

* In this notation, $\ell(J) = {}^{\perp}J$, $\ell^2(J) = {}^{\perp\perp}J$, $\ell^3(J) = {}^{\perp\perp\perp}J$.

This is Lemma 3 of Johns [77]. Now Johns' Lemma 4 (loc. cit.) states that $S = \text{socle } R_R$ is an essential right ideal. To wit: if K is an essential right ideal of R, then $^{\perp}K \subseteq Z \subseteq J$ so $K = (^{\perp}K)^{\perp} \supseteq J^{\perp}$. Since S is the intersection of essential right ideals (see, e.g., Chapter 8 of [73], Kasch-Sandomierski Lemma), this proves that $J^{\perp} \subseteq S$. But $^{\perp}J$ is an essential right ideal,

$$\text{hence } {}^{\perp}J = J^{\perp} \supseteq S,$$

i.e., $S = {}^{\perp}J$.

3.5C Theorem. Any right Noetherian right FGF ring is QF.

Proof. By right FGF, every right ideal is a right annihilator, hence by Lemma 3.5B, R has finite essential right socle, so R is QF by Theorem 3.5.A.

3.6 Theorem (Bjork [72], Tol'skaja [70]). Any right selfinjective right FGF ring Q is QF. Thus, any right PF right FGF ring is QF.*

Proof. Right FGF implies that every simple module embeds in Q, so that Q is an injective cogenerator of mod-Q. By Osofsky's theorem [66], Q must (be semiperfect and) have finite essential right socle. Then Q is QF by 3.5. Since any right PF ring is right selfinjective, the proof is complete.

3.7 Theorem. If every (maximal) left ideal of Q is a left annihilator, and if Q is right FGF, then Q is QF.

* A right PF ring R is defined as an injective cogenerator in mod-R. (See Osofsky [66] or Faith [76-7].)

Proof. Let L be a left ideal, let $I = L^{\perp}$, and let $QA = L_1$ be the left ideal generated by the finite subset $A = \{a_1, \ldots, a_n\}$ such that $A^{\perp} = I$. (We have seen *supra* 2.7 that A exists.) Now L and L_1 are left annulets, so that $L_1 = {}^{\perp}I = L$, proving that every left ideal is finitely generated. Then, (3.3.3) applies.

To obtain the theorem in case every maximal left ideal is a left annihilator, apply Theorem 1 of Kato [68] to see that R_R is injective, so Theorem 3.6 applies. □

In a left cogenerator ring, every left ideal is an annulet, so we have:

3.8 Corollary. *If Q is a cogenerator of Q-mod, and if Q is right FGF, then Q is QF. Thus, any left PF right FGF ring is QF.*

Proof. The first part follows from the theorem. Any left PF ring is left cogenerating, so the second part follows.

4. RIGHT FGF RINGS ARE IF

Right IF rings have been studied in the context of FGF rings by Jain [73]. (Briefly, a ring R is *right IF* if every injective right R-module is flat. Also see Stenstrom [70] and Colby [75].)

For a proof of the next result see Lemma 20 of Rutter [74].

4.1 Theorem (Jain [73a], p. 40, Theorem 3.5 and Rutter [74]). *Any right FGF ring is right IF.*

4.2 Theorem (Jain, loc. cit.).

1. *A right pseudo-coherent right FGF ring is right Noetherian.**

2. *A right FP-injective right FGF ring is right IF and left coherent.*

3. *A right coherent, right FP-injective, right FGF ring is QF.*

4. *A left IF, right FGF ring* R *is right IF and left coherent.*

4.3 Corollary. *A right IF ring* *R* *is QF iff* *R* *has* *$\perp$ acc.*

4.4 Theorem (Jain, loc. cit.). *A right IF ring which is embeddable in a right Noetherian or right perfect ring is right FGF.*†

The proof of the latter depends on a theorem of Simon [72] which asserts that if R can be embedded in a right Noetherian or right perfect ring, then any flat right R-module is an $\aleph_0$-directed union of countably generated modules.

4.5 Corollary. A right Noetherian or right perfect right IF ring is right FGF.†

*Therefore QF by (3.5C).

†Hence, any right FGF ring embeddable in a right Noetherian ring is QF by (3.5C).

5. EMBEDDING TORSIONLESS MODULES IN PROJECTIVES. In this short section, we ask the more natural question: When can a finitely generated torsionless module be embedded in a free module? Bass [60, p. 477, 4.5] shows this happens whenever R is left Noetherian, that is, $Q = Q(R)$ does not have to be Artinian (not to say QF) for this. A module M is torsionless iff there is an embedding of M into a product R^{α} of R, so asking when we can choose the exponent $\underline{\alpha}$ to be finite seems more natural than asking when R is right TF, especially when $R = Q(R)$, the case when every module is torsionfree. (However, when R is a cogenerator of mod-R, then (and only then) every right module is torsionless, so the two problems are equivalent in this case.) The first theorem states that it suffices for the dual module M^* of M to be finitely generated.

5.1 Theorem. *If M is any torsionless right R-module, and if $M^* = \mathrm{Hom}_R(M,R)$ is finitely generated, say $R^n \to M^* \to 0$ exact in R-mod, then $h_R = \mathrm{hom}_R(\ ,R)$ induces an embedding $M \to R^n$ (via an embedding $M^{**} \to R^n$).*

Proof. A module M is torsionless iff the canonical map

$$\begin{cases} M \to R^{M^*} \\ m \to (\quad ,f(m),\ldots) \end{cases}$$

is an embedding. Clearly, if M^* is finitely generated as a left

R-module by $f_1, \ldots, f_n$, then left exactness of h_R converts

$$R^n \to M^* \to 0$$

exact into

$$0 \to M^{**} \to R^n$$

exact. Then, M torsionless means $M \hookrightarrow M^{**}$ canonically, so $M \to R^n$. □

5.2 Theorem. 1. _If_ R _right selfinjective, then_ h_R _converts any embedding_ $M \hookrightarrow R^n$ _into an exact sequence_ $R^n \to M^* \to 0$ _in_ R-mod.

2. (Converse of 1.) _If every embedding_ $M \to R^n$ is _converted by_ h_R _into a canonical exact sequence_ $R^n \to M^* \to 0$, _then_ R _is right selfinjective._

Proof. 1. is trivial since h_R is exact. Also 2. follows from Baer's criterion for injectivity of the right module R. □

5.3 Corollary. 1. _If_ h_R : mod-R $\rightsquigarrow$ mod-R _preserves finitely generated modules, then every finitely generated torsionless right_ R-_module embeds in a free module_.

2. _If_ R _is a right cogenerator, and if_ h_R : mod-R $\rightsquigarrow$ R-mod _preserves finitely generated modules, then_ R _is right FGF_.

3. _If_ R _is a right injective cogenerator_, (i.e. _right PF_) _then a right_ R _module_ M _embeds in a finite free_ R-_module iff_ M^* _is finitely generated_.

Proof. 1. follows from Theorem 5.1, inasmuch as M is torsionless, and M^* is f.g. Then 2. follows, since R is a right cogenerator iff every right module is torsionless. 3. If M^* is f.g. by n elements, then $M \rightarrow R^n$ by Theorem 5.1. Conversely. □

Note in 5.1, M is not assumed to be finitely generated. If it is, then $R^n \rightarrow M \rightarrow 0$ exact in mod-R, implies $0 \longrightarrow M^* \longrightarrow R^n$ exact in R-mod, which shows that M^* is finitely generated when R is left Noetherian. (Thus, as stated, $M \hookrightarrow R^n$ in this case.) This suffices to prove Bass' result referred to earlier.

The proof of this remark shows that _the dual module_ M^* _of a finitely generated right_ R-_module_ M _embeds in a finitely generated left free module._

ACKNOWLEDGEMENT

This paper was discussed impromptu during the problem session.

I sketched the proofs of several results; Professor M. Auslander gave an alternative proof of Theorem 3.5C.

REFERENCES

[63] Levy, L.S., Torsion free and divisible modules over non-integral-domains, Can. J. Math. 15 (1963) 132-151.

[66] Osofsky, B.L., A generalization of Quasi-Frobenius rings, J. Algebra 4 (1966) 373-387.

[67] Faith, C. and Walker, E.A., Direct sum representations of injective modules, J. Algebra 5 (1967) 203-221.

[68] Kato, T., Torsionless modules, Tohoku Math. J. 20 (1968) 234-243.

[70] Tol'skaja, T.S., When are all cyclic modules essentially embedded in free modules, Mat. Issled 5 (1970) 187-192.

[70] Stenstrom, B., Coherent rings and FP-injective modules, J. London Math. Soc. 2(2) (1970) 323-329.

[72] Bjork, J.E., Radical properties of perfect modules, J. Reine u. Angew. Math., 245 (1972) 78-86.

[72] Simon, D., On the structure of flat modules, Bull. de L'Academie Polanaise des Sciences, 20(2) (1972) 115-120.

[73] Faith, C., Algebra: Rings, Modules, and Categories, Grundlehren der Math. Wiss., Bd. 190, Springer-Verlag, New York and Berlin, 1973; corrected reprint 1981.

[73a] Jain, S., Flat injective modules and FP-injectivity, Ph.D. Thesis, Rutgers U., 1973, New Brunswick, N.J. 08903.

[73b] Jain, S., Flat and FP-injectivity, Proc. Amer. Math. Soc. (1973).

[74] Rutter, E.A., A characterization of QF-3 rings, Pac. J. Math 51 (1974) 533-6.

[75] Colby, Rings which have flat injectives, J. Algebra 35 (1975) 239-52.

[76-7] Faith, C., Injective cogenerator rings and a theorem of Tachikawa. I,II, Proc. Amer. Math. Soc. 60 (1976) 25-30; 62 (1977) 15-18.

[77] Johns, B., Annihilator conditions in Noetherian rings, J. Algebra 5 (1967) 203-221.

[81] Menal, P., On the endomorphism ring of a free module, preprint, Seccio de Matematiques, Universitat Autonomo de Barcelona, 1981.

MAXIMAL TORSION RADICALS OVER RINGS WITH FINITE REDUCED RANK

John A. Beachy
Northern Illinois University
DeKalb, Illinois

For a left Noetherian ring R with prime radical N, the reduced rank of R has been defined by Goldie [9] as follows. Let Q be the semisimple classical ring of left quotients of R/N, and assume that k is the index of nilpotence of N. Then the reduced rank of R, denoted by $\rho(R)$, is defined by the formula

$$\rho(R) = \sum_{i=1}^{k} \ell(Q \otimes_R (N^{i-1}/N^i)),$$

where $\ell(X)$ denoted the length of a Q-module X.

Let γ denote the torsion radical of R-Mod cogenerated by E(R/N), with the associated quotient functor denoted by Q_γ: R-Mod $\to$ R-Mod/γ. If R is left Noetherian, then it follows from a result of Jategaonkar [11, Proposition 1.9] that $Q_\gamma(R)$ has finite length as an object in the quotient category R-Mod/γ. This length is precisely the reduced rank of R [7, Proposition 2]. Thus it is possible to give the following more general definition. The ring R is said to have *finite reduced rank* (on the left) if $Q_\gamma(R)$ has finite length in the quotient category R-Mod/γ cogenerated by E(R/N).

Lenagan [12] has shown that Goldie's definition of reduced rank can be extended to any ring with Krull dimension. It also follows from [6, Proposition 6] that if R has Krull dimension, then it has finite reduced rank in the general sense. Furthermore, [7, Theorem 4] shows that the ring R is a left order in a left Artinian ring if and only if R has finite reduced rank (on the left) and satisfies the regularity condition. Thus the class of rings with finite reduced rank is quite extensive. In Theorem 4 below it will be shown that if R has finite reduced rank and S is Morita equivalent to R, then S also has finite reduced rank.

The Walkers [15, Theorem 1.29] showed that for a commutative ring R there is a one-to-one correspondence between maximal torsion radicals of R-Mod and minimal prime ideals of R. This correspondence holds for any ring with Krull dimension by [4, Theorem 4.6], and furthermore, in this case every torsion radical is contained in a maximal torsion radical. Theorem 2 below shows that these conditions hold for any ring with finite reduced rank whose prime radical is right T-nilpotent. In fact, rings with finite reduced rank can be characterized by conditions involving maximal torsion radicals (see Theorem 1).

Throughout the paper, R will be assumed to be an associative ring with identity element, and all modules will be assumed to be unital R-modules. The category of left R-modules will be denoted by R-Mod, the direct sum of n isomorphic copies of a module $_RM$ will be denoted by M^n, and the injective envelope of M will be denoted by $E(M)$. The reader is referred to the book by Stenström [14] for definitions and results on quotient categories and torsion radicals.

If σ is a torsion radical (called a left exact radical by Stenström), then the quotient category it determines will be denoted by R-Mod/σ, with the exact quotient functor denoted by Q_σ: R-Mod $\to$ R-Mod/σ. A module $_RM$ is called σ-torsionfree if $\sigma(M) = (0)$, and σ-torsion if $\sigma(M) = M$; a submodule $M' \subseteq M$ is called σ-closed if M/M' is σ-torsionfree, and σ-dense if M/M' is σ-torsion. Recall that the subobjects of $Q_\sigma(M)$ in R-Mod/σ correspond to the σ-closed submodules of M [14, Chapter IX, Corollary 4.4]. Thus in the quotient category R-Mod/σ, the object $Q_\sigma(M)$ has finite length if and only if M satisfies both ascending and descending chain conditions on the set of σ-closed submodules. In this case, $M/\sigma(M)$ must have finite uniform dimension.

If $_RX$ is an injective module, then X defines a torsion radical rad_X by setting

$$\mathrm{rad}_X(M) = \{m \in M \mid f(m) = 0 \quad \text{for all} \quad f \in \mathrm{Hom}_R(M,X)\}$$

for any module M in R-Mod. If $\sigma = \mathrm{rad}_X$, then $X = Q_\sigma(X)$ is a cogenerator for R-Mod/σ, and σ is said to be the torsion radical cogenerated by X.

The torsion radical σ is said to be *prime* if $\sigma = \mathrm{rad}_{E(M)}$ for a module $_RM$ such that $Q_\sigma(M)$ is a simple object in R-Mod/σ. Equivalently, R-Mod/σ has a cogenerator which is the injective envelope of a simple object. In this case, any cogenerator of R-Mod/σ contains an isomorphic copy of the simple object. Note that if P is a prime Goldie ideal of R (that is, R/P is a prime left Goldie ring), then $\mathrm{rad}_{E(R/P)}$ is a prime torsion radical.

If σ and τ are torsion radicals, then $\sigma \le \tau$ if $\sigma(M) \subseteq \tau(M)$ for each module M in R-Mod. Note that $\sigma = \mathrm{rad}_{E(X)}$ is the largest torsion radical such that X is a σ-torsion. If σ is a proper torsion radical (that is, σ is not the identity functor on R-Mod), then σ is said to be *maximal* if $\sigma \le \tau$ implies $\sigma = \tau$ for all proper torsion radicals τ. By [3, Theorem 2.4], σ is maximal if and only if the quotient cateogry R-Mod/σ is nonzero and each nonzero σ-torsionfree injective module is a cogenerator of R-Mod/σ.

THEOREM (1). *The following conditions are equivalent.*

(1) The ring R *has finite reduced rank (on the left).*

(2) There exist maximal torsion radicals $\mu_1,\ldots,\mu_n$ *such that the localization of* R *has finite length in* R-Mod/μ_i, *for* $i = 1,\ldots,n$, *and such that for each prime ideal* P *of* R, $\mathrm{rad}_{E(R/P)} \le \mu_i$ *for some* i.

Proof. (1) => (2). Let N be the prime radical of R, and let $\gamma = \mathrm{rad}_{E(R/N)}$. By assumption $Q_\gamma(R)$ has finite length in R-Mod/γ, so R/N is a semiprime left Goldie ring by [6, Proposition 2]. Let P be a minimal prime ideal of R, and let

$\mu = \mathrm{rad}_{E(R/P)}$. Then $\gamma \le \mu$, so every μ-closed left ideal of R is γ-closed, which implies that R/P is a left Goldie ring, and so μ is a prime torsion radical. Furthermore, $Q_\mu(R)$ must have finite length in $R\text{-Mod}/\mu$, and so each μ-torsionfree injective module contains a submodule which maps to a simple object in the quotient category. This shows that any μ-torsionfree injective module cogenerates $R\text{-Mod}/\mu$, and therefore μ is a maximal torsion radical.

Let $P_1,\ldots,P_n$ be the minimal prime ideals of R, and let μ_i be the torsion radical cogenerated by $E(R/P_i)$. If P is any prime ideal of R, then $P \supseteq P_i$ for some i, and so $\mathrm{rad}_{E(R/P)} \le \mu_i$ by [3, Lemma 2.5] since P_i is a prime Goldie ideal.

(2) $\Longrightarrow$ (1). Assume that condition (2) holds. If R has finite length with respect to μ_i, then by [6, Theorem 7], $\mu_i = \mathrm{rad}_X$ for an injective module $X \cong \oplus_{\alpha\in I} E(R/P_\alpha)$, where each ideal P_α is a prime Goldie ideal. If P_i is any one of the prime ideals in the decomposition of X, then $E(R/P_i)$ cogenerates a torsion radical larger than or equal to μ_i, so the maximality of μ_i implies that $P_\alpha = P_i$ for all $\alpha \in I$.

If Q is a minimal prime ideal of R, then $\mathrm{rad}_{E(R/Q)} \le \mu_i$ for some i, which implies that $P_i \subseteq Q$ by [3, Lemma 3.5], and so $P_i = Q$. Thus each minimal prime ideal is represented in the set $P_1,\ldots,P_n$, and each minimal prime ideal is a prime Goldie ideal since R has finite length with respect to μ_i. If P_i were not a minimal prime for some i, then $P_i \supseteq P_j$ for some minimal prime P_j with $j \ne i$. The maximality of μ_i would then imply that $P_i = P_j$, a contradiction. Thus $P_1,\ldots,P_n$ is the set of minimal prime ideals of R. Since the prime torsion radicals $\mu_1,\ldots,\mu_n$ are incomparable and R has finite length with respect to μ_i, for each i, it follows from [10, Theorem 3.6] that R has finite length with respect to $\mathrm{rad}_{E(R/N)} = \cap_{i=1}^{n}\mu_i$, and thus R has finite reduced rank. □

An ideal I of R is said to be right *T-nilpotent* in case for each sequence $a_1, a_2, \ldots$ of elements in I there exists an integer n such that $a_n \cdots a_2 a_1 = 0$. The following example shows that the prime radical of a ring with finite reduced rank may be T-nilpotent but not nilpotent. (Example 2 of [7] shows that the prime radical need not even be T-nilpotent.)

Example. Let $R = F[x_1, x_2, \ldots]$ be the ring of polynomials in infinitely many indeterminates over a field F, and let I be the ideal generated by $x_2^2, x_3^3, \ldots$ and the products $x_i x_j$, for $i \ne j$. The ideal N generated by $x_2, x_3, \ldots$ is a prime ideal, and since each element of N is nilpotent in R/I, it follows that N/I is the prime radical of R/I. Furthermore, $x_1 \notin N$ but $x_1 x_j \in I$ for $j \ne 1$, and so N/I is torsion with respect to the torsion radical γ of $R/I\text{-Mod}$ cogenerated by $E_{R/I}(R/N)$. Thus the localization $Q_\gamma(R/I)$ is just the quotient field of R/N, and $\rho(R/I) = 1$. It can be immediately checked that N/I is T-nilpotent.

Thus R/I is a ring with finite reduced rank whose prime radical is T-nilpotent but not nilpotent.

THEOREM (2). *The following conditions are equivalent for a ring* R *with prime radical* N.

(1) *The ring* R *has finite reduced rank (on the left) and* N *is right* T-*nilpotent.*

(2) *(i)* *The category* R-Mod *has maximal torsion radicals* $\mu_1,\dots,\mu_n$ *such that for* $i = 1,\dots,n$ *the ring* R *has finite length with respect to* μ_i*;*

(ii) *for each torsion radical* σ *of* R-Mod, $\sigma \le \mu_i$ *for some* i*,* $1 \le i \le n$;

(iii) *the maximal torsion radicals of* R-Mod *correspond to minimal prime ideals of* R.

Proof. (1) $\Rightarrow$ (2). Condition (i) follows from Theorem 1, which also shows that R/N is a left Goldie ring. By [5, Theorem 1.8], conditions (ii) and (iii) hold for R/N, and then they hold for R by [5, Proposition 2.7] since N is right T-nilpotent.

(2) $\Rightarrow$ (1). By [5, Proposition 2.7], conditions (ii) and (iii) imply that N is right T-nilpotent. Conditions (i) and (ii) imply that R has finite reduced rank, by Theorem 1. □

Goldman has shown [10, Theorem 5.10] that if R is a left Noethrian ring, then R is left Artinian if and only if every prime torsion radical is maximal. The next theorem extends this result to certain rings with finite reduced rank.

THEOREM (3). *Let* R *be a ring with finite reduced rank (on the left) such that the prime radical of* R *is right* T-*nilpotent. Then* R *is left Artinian if and only if every prime torsion radical of* R-Mod *is maximal.*

Proof. If R is left Artinian, then every prime torsion radical is maximal by [10, Theorem 5.10]. Conversely, if every prime torsion radical is maximal, then each simple module defines a maximal torsion radical. By Theorem 2, there are only finitely many maximal torsion radicals, and so R-Mod has only finitely many nonisomorphic simple modules $S_1,\dots,S_n$. If the torsion radical μ_i is cogenerated by $E(S_i)$, $i = 1,\dots,n$ then ${}_RR$ has finite length with respect to $\mu = \cap_{i=1}^n \mu_i$. Since $S_1,\dots,S_n$ is a complete set of representatives of the equivalence classes of simple modules, $\oplus_{i=1}^n E(S_i)$ is a cogenerator for R-Mod, which implies that μ is the zero functor. This shows that ${}_RR$ has finite length. □

THEOREM (4). *Let* R *and* S *be Morita equivalent rings. If* R *has finite reduced rank (on the left), then so does* S.

Proof. The proof will make use of the characterization of rings with finite reduced rank given in Theorem 1. Let F: R-Mod $\rightarrow$ S-Mod be an equivalence of categories.

If ${}_RX$ is cogenerated by ${}_RY$, then by [1, Proposition 21.6], ${}_SF(X)$ is cogenerated by ${}_SF(Y)$. Since F preserves injective modules, this shows that F preserves the lattice of torsion radicals. Furthermore, corresponding torsion radicals determine isomorphic quotient categories, and so S must have finitely many maximal torsion radicals, and it must have finite length with respect to each of these. The proof can be completed by showing that F preserves torsion radicals of the form $rad_{E(R/P)}$, where P is a prime ideal.

As shown by [1, Proposition 21.11], ideals of R corresponds to ideals of S by assigning to the ideal I of R the ideal $Ann_S(F(R/I))$ of S. The characterization of prime ideals given in [2, Theorem 2] shows that if P is a prime ideal of R, then the R-module R/P has the property that if ${}_RX$ is cogenerated by R/P, then R/P is cogenerated by X. Furthermore, if ${}_SY$ has the property that ${}_SY$ is cogenerated by ${}_SX$ whenever X is cogenerated Y, then $Ann_S(Y)$ is a prime ideal. This shows that under the one-to-one correspondence between ideals of R and S defined above, if P is a prime ideal of R, then there is a prime ideal Q of S such that F(R/P) cogenerates and is cogenerated by S/Q. It follows that E(S/Q) and $F(E(R/P)) \simeq E(F(R/P))$ cogenerate the same torsion radical, so that the equivalence between R-Mod and S-Mod preserves the torsion radicals corresponding to prime ideals. □

References

[1] F. W. Anderson and K. R. Fuller, Rings and Categories of Modules, Springer-Verlag, New York/Heidelberg/Berlin, 1974.

[2] J. A. Beachy, A characterization of prime ideals, J. Indian Math. Soc. 37(1973), 343-345.

[3] ___________, On maximal torsion radicals, Canadian J. Math. 25(1973), 712-726.

[4] ___________, On maximal torsion radicals II, Canadian J. Math. 27(1975), 115-120.

[5] ___________, Some aspects of noncommutative localization, Noncommutative Ring Theory, Kent State, 1975, Lecture Notes in Mathematics #545, pp. 2-31.

[6] ___________, Injective modules with both ascending and descending chain conditions on annihilators, Commun. Algebra 6(1978), 1777-1788.

[7] ___________, Rings with finite reduced rank, Commun. Algebra (to appear).

[8] A. W. Chatters, A. W. Goldie, C. R. Hajarnavis and T. H. Lenagan, Reduced rank in Noetherian rings, J. Algebra 61(1979), 582-589.

[9] A. W. Goldie, Torsion-free modules and rings, J. Algebra 1(1964), 268-287.

[10] O. Goldman, Elements of noncommutative arithmetic I, J. Algebra 35(1975), 308-341.

[11] A. V. Jategaonkar, Relative Krull dimension and prime ideals in right Noetherian rings, Commun. Algebra 4(1974), 429-468.

[12] T. H. Lenagan, Reduced rank in rings with Krull dimension, Ring Theory, Proceedings of the 1978 Antwerp Conference, Marcel Dekker, Inc., New York/Basel, 1979, 123-131.

[13] R. W. Miller and M. Teply, The descending chain condition relative to a torsion theory, Pacific J. Math. 83(1979), 207-219.

[14] B. Stenström, Rings of Quotients, Springer-Verlag, New York/Heidelberg/Berlin, 1975.

[15] C. L. Walker and E. A. Walker, Quotient categories and rings of quotients, Rocky Mountain J. Math. 2(1972), 513-555.

STABLE RINGS WITH FINITE GLOBAL DIMENSION

by

Robert F. Damiano and Zoltan Papp

Department of Mathematics
George Mason University
Fairfax, Virginia 22030

1. Background and Notation. By a ring R, we shall always mean an associative ring with unit. We say a ring R is *left stable* [9] if for any injective left R-module H and for any left R-module M, the relation $\mathrm{Hom}_R(M,H) = 0$ implies that $\mathrm{Hom}_R(E(M),H) = 0$ where E(M) is the injective envelope of M. A ring is *stable* if it is right and left stable. Similarly, a ring is noetherian if it is right and left noetherian.

A left noetherian ring R is *left bounded* if every essential left ideal of R contains a nonzero two-sided ideal. A left noetherian ring R is said to be fully left bounded (FLBN), if R/P is left bounded for every prime ideal P. If R is both right and left noetherian and both right and left fully bounded, we say R is an FBN ring. A special subfamily of FBN rings are those noetherian rings satisfying a polynomial identity (PI rings).

Over a FLBN ring, left stability is equivalent to the condition that each prime ideal P satisfies the *left Artin-Rees property*, i.e., if I is a left ideal of the ring R, then there exists an $n \in \mathbb{N}$ such that $P^n \cap I \subseteq PI$ [10, Proposition, p. 310].

We say P satisfies the <u>Artin-Rees property</u> if it satisfies the left and right Artin-Rees property. The Krull dimension of a left R-module ${}_RM$, denoted by K-dim(${}_RM$), is the dimension introduced by Rentschler and Gabriel. An exposition on the properties of Krull dimension can be found in [12]. In particular, every finitely generated left R-module over a noetherian ring has Krull dimension [12, Prop. 1.3]. Moreover, for an FBN ring R, the left Krull dimension of R equals the classical Krull dimension derived from chains of prime ideals [18, Thm. 2.4]. By symmetry, for an FBN ring R, the left Krull dimension of R equals the right Krull dimension of R.

A left noetherian has a perfect left localization R ring at a prime ideal P if $C(P) = \{r \in R \mid r + P \text{ is regular in } R/P\}$ has the left Öre condition [29, Prop. II. 1.5]. We denote this localization by R_P. Although left stability in a noetherian ring is not enough to insure perfect left localizability, Smith [27, Prop. 3.4] has proved that a noetherian stable ring R has a perfect localization (left and right) at every prime ideal P. Another case in which R has a perfect localization at every prime ideal is given by the following proposition.

<u>Proposition 1</u>: A right noetherian, left stable FLBN ring R can be left localized at every prime ideal P. Moreover R_P is a left stable FLBN ring.

<u>Proof</u>: By [19, Cor. 12], P is ideal invariant, i.e. for every left ideal I, K-dim(P/PI) $\leq$ K-dim(R/I). Since R is left stable, P satisfies the left Artin-Rees property. By [23, Prop. 9]

R has a left perfect localization at P. R_P is left noetherian [10, Prop. 17.13]. Suppose $\mathrm{Hom}_{R_P}(N,H) = 0$ where H is a left injective R_P-module. By [10, Prop. 17.8], H is a left injective R-module and $\mathrm{Hom}_{R_P}(N,H) = \mathrm{Hom}_R(N,H)$ [26, Lemma 11.61]. Since R is left stable, $\mathrm{Hom}_R(E(N),H) = 0$. Again by [26, Lemma 11.61], $\mathrm{Hom}_{R_P}((E(N)_P,H) = 0$. By [6, Prop. 1], $(E(N))_P$ is the injective envelope of N. Therefore R_P is a left stable ring. It is well known that R is FLBN if and only if there exists a 1-1 correspondence between the prime ideals and the indecomposable injective R-modules [29, Thm. VII. 2.1]. Since left localization at P is a perfect localization, a 1-1 correspondence exists between the prime ideals of R_P and the indecomposable injective R_P-modules [10, Prop. 17.14]. Therefore R_P is FLBN. □

We note that we know of no example of a ring R satisfying the hypotheses of proposition 1 where R_P is not also right noetherian. If R is stable FBN, then we get a much stronger result.

<u>Corollary 2</u>: If R is a stable FBN ring with P a prime ideal then R_P is a stable FBN ring.

<u>Proof</u>: Müller [24, Thm. 11] has shown that for stable FBN rings, the left localization of R at P coincides with the right localization of R at P. The result now follows from proposition 1. □

It is of interest to remark that if right noetherian is

omitted from the hypotheses of proposition 1, R may not be left localizable at every prime ideal.

Example: Let $S = F[[x]]$, the power series ring over a commutative field F. Let $M = F[[x]]$ viewed naturally as a left S-module, but as a right $F[[x]]$-module via the quotient ring epimorphism

$$F[[x]] \to F \to 0.$$

One can show that the trivial extension of S by M, $R = \{\begin{pmatrix} s & 0 \\ m & s \end{pmatrix} \mid s \in S,\ m \in M\}$, is a left noetherian, left stable PI ring. However, R is not right noetherian. The prime ideal $P = \{\begin{pmatrix} 0 & 0 \\ m & 0 \end{pmatrix} \mid m \in M\}$ satisfies the Artin-Rees property but $\mathcal{C}(P)$ does not satisfy the left Öre condition. Thus R is not left localizable at P. □

The Krull dimension of a left R-module M will be denoted by K-dim(M). The injective (projective) dimension of M will be denoted by inj dim(M)(proj dim(M)) while the global dimension of R will be denoted by gl dim(R). For definitions of Krull, global, injective, and projective dimensions we refer the reader to [12, 26].

Finally, we shall call a noetherian ring R local if it has a unique maximal ideal J such that R/J is a division ring. A noetherian ring R shall be called (left) classically local if R has a unique maximal ideal J such that R/J is a simple artinian ring and J satisfies the (left) Artin-Rees property.

2. Global Dimension of Stable FBN Rings. The Krull dimension of a commutative noetherian ring R is always less than or equal to the global dimension of R [3, Lemma 3.1]. Furthermore for commutative noetherian rings having finite global dimension, K-dim(R) = gl dim(R). We shall show that this result can be generalized to stable FBN rings. We should note that the above relationship between Krull dimension and global dimension is not true for all left noetherian rings [12, Example 10.3]. However, the question is open for noetherian rings in general. The reader can find some further information in the following sources [5, 17, 25, 31]. In general, it is not true for noetherian rings that if a ring has finite global dimension, the Krull dimension equals the global dimension. An example is the ring of 2 × 2 upper triangular matrices over a division ring.

To attain the result, we first need some propositions on the structure of Ext. Recall that for a commutative noetherian ring R, if M and N are finitely generated R-modules, then $\mathrm{Ext}_R^k(M,N)$ is a finitely generated R-module. This is a key element in the commutative proofs [3]. However, if ${}_RM$ and ${}_RN$ are left R-modules, $\mathrm{Ext}_R^k(M,N)$ need not even have an R-module structure.

Proposition 3: Let R be a classically local noetherian ring with unique maximal ideal J. If ${}_RM$ is a left R-module, then $\mathrm{Ext}_R^k(R/J,M) \cong \mathrm{Soc}(E_k)$, where E_k is the kth-injective module in the minimal injective resolution of M.

Proof: Let

$$0 \to M \to E_0 \xrightarrow{d_0} E_1 \xrightarrow{d_1} \cdots \to E_{k-1} \xrightarrow{d_{k-1}} E_k \xrightarrow{d_k} E_{k+1} \to \cdots$$

be a minimal injective resolution of M. Let $T_k = \mathrm{Hom}_R(R/J, E_k)$ and d_k^*, the induced R-homomorphism

$$d_k^* : T_k \to T_{k+1} .$$

We claim that $d_k^*(T_k) = 0$. Suppose $0 \neq x \in T_k$. Therefore, $Jx = 0$. Since R is classically local, $Rx = \oplus S_\alpha$ where S_α are isomorphic copies of the unique simple module. But $E_k = E_k(\ker d_k)$ the injective envelope of $\ker d_k$. Thus, $\oplus S_\alpha \leq \ker d_k^*$. Therefore $T_k = \ker d_k^*$ and hence $\mathrm{Im}\, d_{k-1}^* = 0$. Since $\mathrm{Ext}_R^k(R/J, M) = \ker d_k^*/\mathrm{Im}\, d_{k-1}^* = \ker d_k^* = T_k \cong \mathrm{Soc}(E_k)$, we are done. □

Although the following proposition is probably well-known, we were unable to locate a reference for it. For the convenience of the reader, we shall give a proof for it.

Proposition 4: Let R be a left noetherian ring. Let $\{N_\alpha \mid \alpha \in A\}$ be a directed family of left R-modules. If M is a finitely generated left R-module, then

$$\mathrm{Ext}_R^k\,(M, \varinjlim N_\alpha) \cong \varinjlim \mathrm{Ext}_R^k(M, N_\alpha)$$

Proof: We prove the result by induction on k. Let

$$0 \to K \to P \to M \to 0$$

be an exact sequence with P, a finitely generated projective R-module. Since R is left noetherian, K is also left finitely generated. The diagram

$$\begin{array}{ccccccc}
\mathrm{Hom}_R(P,\varinjlim N_\alpha) & \to & \mathrm{Hom}_R(K,\varinjlim N_\alpha) & \to & \mathrm{Ext}_R^1(M,\varinjlim N_\alpha) & \to & 0 \\
\downarrow \alpha & & \downarrow \beta & & \downarrow \gamma & & \\
\varinjlim \mathrm{Hom}_R(P,N_\alpha) & \to & \varinjlim \mathrm{Hom}_R(K,N_\alpha) & \to & \varinjlim \mathrm{Ext}_R^1(M,N_\alpha) & \to & 0
\end{array}$$

is exact with α and β being isomorphisms [29, Prop. 3.4]. Thus the result follows for $k = 1$. If the result is valid for all natural numbers less than or equal to $k-1$, then

$$\begin{aligned}
\mathrm{Ext}_R^k(M,\varinjlim N_\alpha) &\cong \mathrm{Ext}_R^{k-1}(K,\varinjlim N_\alpha) \\
&\cong \varinjlim \mathrm{Ext}_R^{k-1}(K,N_\alpha) \\
&\cong \varinjlim \mathrm{Ext}_R^k(M,N_\alpha)
\end{aligned}$$
□

Proposition 5: Let R be a left classically local noetherian ring with unique simple module S. If M is a left finitely generated R-module such that $\mathrm{Ext}_R^k(M,R/J) = 0$, then $\mathrm{Ext}_R^k(M,N) = 0$ for all submodules $N \subseteq E(S)$.

Proof: Suppose $\mathrm{Ext}_R^k(M,R/J) = 0$. Since R/J is a direct sum of simple modules, each isomorphic to S, $\mathrm{Ext}_R^k(M,S) = 0$. By induction on length, it follows that $\mathrm{Ext}_R^k(M,N) = 0$ for all modules N of finite length. Suppose $N \subseteq E(S)$, then $N = \varinjlim N_\alpha$ where each N_α has finite length [13, Prop. 4.3]. By proposition 4, $\mathrm{Ext}_R^k(M,N) \cong \varinjlim \mathrm{Ext}_R^k(M,N_\alpha) = 0$.

Proposition 6: Let R be a left noetherian ring with I an ideal and T a right noetherian ring. If M is an (R,T)-bimodule, finitely generated as a right T-module, then $\mathrm{Ext}_R^k(R/I,M)$ is an (R/I,T)-bimodule, finitely generated as a right T-module.

Proof: Let

$$\cdots \to P_{k+1} \xrightarrow{d_k} P_k \xrightarrow{d_{k-1}} P_{k-1} \to \cdots \to P_1 \xrightarrow{d_0} P_0 \to R/I \to 0.$$

be a finitely generated projective resolution of R/I. Applying the functor $\mathrm{Hom}_R(-,{}_RM_T)$, we get the derived sequence

$$0 \to \mathrm{Hom}_R(P_0,M) \xrightarrow{d_0^*} \mathrm{Hom}_R(P_1,M) \to \cdots \to \mathrm{Hom}_R(P_{k-1},M) \xrightarrow{d_{k-1}^*} \mathrm{Hom}_R(P_k,M) \xrightarrow{d_k^*} \cdots$$

Each $\mathrm{Hom}_R(P_i,M)$ is a finitely generated right T-module since M is a finitely generated right T-module. One can easily check that each d_i^* is a right T-homomorphism. Since T is right noetherian, $\mathrm{Ext}_R^k(R/I,M) = \ker d_k^*/\mathrm{im}\, d_{k-1}^*$ is finitely generated as a right T-module. The left R/I structure comes from applying the functor $\mathrm{Hom}_R(R/I,-)$ to an injective resolution of M. □

Theorem 7: If R is a stable FBN ring, then $\text{K-dim}(R) \leq \text{gl dim}(R)$ If, in addition, R has finite global dimension, then $\text{K-dim}(R) = \text{gl dim}(R)$.

Proof: By [6, Prop. 1, Cor. 3, and Cor. 2], it suffices to assume R has been localized at a maximal ideal. Let J be the unique maximal ideal of R. We note that since R is a stable ring, J satisfies the Artin-Rees property. By [14, Cor. 3.8], the Krull dimension of a semilocal FBN ring is finite. Therefore, without loss of generality, assume the global dimension of R is finite. We will prove the results by induction on the global dimension.

If $\text{gl dim}(R) = 0$, then R is semisimple and we are done. Suppose the result is true for global dimensions less than or equal to k-1 and let R be a stable FBN ring with global dimension k and unique maximal ideal J. Let P be a prime ideal of R such that $K\text{-dim}(R_P) = K\text{-dim}(R)-1$. By Goldie's theorem [29, Thm. II. 2.2], $J \cap C(P) \neq \emptyset$. Let $x \in J \cap C(P)$. Consider the exact sequence

$$0 \to R/P \overset{x}{\to} R/P \to R/Rx+P \to 0$$

where the monomorphism is given by right multiplication by x.

This sequence gives rise to the long exact sequence

$$\ldots \to \mathrm{Ext}_R^{k-1}(R/P,R/P) \to \mathrm{Ext}_R^{k-1}(R/P,R/Rx+P) \to \mathrm{Ext}_R^{k}(R/P,R/P) \overset{x}{\to}$$
$$\mathrm{Ext}_R^{k}(R/P,R/P) \to \mathrm{Ext}_R^{k}(R/P,R/Rx+P) \to 0 \qquad (1)$$

where the map from $\mathrm{Ext}_R^k(R/P,R/P)$ to $\mathrm{Ext}_R^k(R/P,R/P)$ is given by right multiplication by x.

We first claim that $\mathrm{Ext}_R^k(R/P,R/J) = 0$. Suppose $\mathrm{Ext}_R^k(R/P,R/J) \neq 0$. Since R is stable with unique simple submodule S, the minimal injective resolution of R/J has the property that $E_j = \oplus E(S)$ for all j. Let $0 \neq \bar{\varphi} \in \mathrm{Ext}_R^k(R/P,R/J) = \ker d_k^*/\mathrm{Im}\, d_{k-1}^*$. Pick $\varphi \in \mathrm{Hom}_R(R/P,E_k)$ such that $\varphi \in \ker d_k^*$ and $\varphi + \mathrm{Im}\, d_{k-1}^* = \bar{\varphi}$. Since $E_k = \oplus E(S)$ and J has the Artin-Rees property, $R\varphi(R/P)$ has finite length [13, Prop. 4.3]. In particular, there exists a power of x such that $x^\ell\varphi(R/P) = 0$. Consider the exact sequence

$$0 \to R/P \xrightarrow{x^{\ell+1}} R/P \to R/Rx^{\ell+1}+P \to 0$$

The functor Ext (-,R/J) gives a long exact sequence

$$\ldots \to \mathrm{Ext}_R^k(R/Rx^{\ell+1}+P,R/J) \to \mathrm{Ext}_R^k(R/P,R/J) \xrightarrow{x^{\ell+1}} \mathrm{Ext}_R^k(R/P,R/J) \to 0$$

where the last map is given by left multiplication by $x^{\ell+1}$. Since by proposition 6, $\mathrm{Ext}_R^k(R/P,R/J)$ is a right finitely generated R-module, left multiplication by $x^{\ell+1}$ is an isomorphism [7, Thm. 1]. But $x^{\ell+1} \cdot \varphi(r) + \mathrm{Im}\, d_{k-1}^* = \varphi(rx^{\ell+1}) + \mathrm{Im}\, d_{k-1}^* = rx^{\ell}(\varphi(x)) + \mathrm{Im}\, d_{k-1}^* = 0$. Since $0 \neq \bar{\varphi} \in \mathrm{Ext}_R^k(R/P,R/J)$, we get a contradiction. Therefore, $\mathrm{Ext}_R^k(R/P,R/J) = 0$.

We selected P so that K-dim(R/P) = 1. Hence, K-dim (R/Rx+P) = 0 [12, Prop. 6.2]. Thus R/Rx+P is a left R-module with finite length.

Therefore by proposition 5, $\mathrm{Ext}_R^k(R/P,R/Rx+P) = 0$. Our long exact sequence (1) now becomes

$$\cdots \to \mathrm{Ext}_R^{k-1}(R/P,R/P) \to \mathrm{Ext}_R^{k-1}(R/P,R/Rx+P) \to \mathrm{Ext}_R^k(R/P,R/P) \xrightarrow{x} \mathrm{Ext}_R^k(R/P,R/P) \to 0.$$

But $\mathrm{Ext}_R^k(R/P,R/P)$ is a right finitely generated R/P-module by proposition 6 and x is a nonzero element of the Jacobson radical of R/P. By Nakayama's Lemma [1,Cor. 15.13], $\mathrm{Ext}_R^k(R/P,R/P) = 0$.

Let $R_P \otimes R/P = (R/P)_P$. Consider the exact sequence

$$0 \to R/P \to (R/P)_P \to K \to 0$$

where K is the cokernel of the inclusion map from R/P to $(R/P)_P$. Since $R_P \otimes K = 0$ and since R is stable, $K = \varinjlim K_\alpha$ where each K_α has finite length [13, Prop. 4.3]. Applying the functor Ext(R/P,-), we have the long exact sequence

$$\cdots \to \mathrm{Ext}_R^k(R/P,R/P) \to \mathrm{Ext}_R^k(R/P,(R/P)_P) \to \mathrm{Ext}_R^k(R/P,K) \to 0$$

Since $\mathrm{Ext}_R^k(R/P,R/P) = 0$ and $\mathrm{Ext}_R^k(R/P,K) = 0$ by proposition 5, $\mathrm{Ext}_R^k(R/P,(R/P)_P) = 0$. By [26, Thm. 11.65]

$Ext^k_{R_P}((R/P)_P,(R/P)_P) \cong Ext^k_R(R/P,(R/P)_P) = 0$. Boratynski's result [4, Lemma 3] gives gl $\dim(R_P) = $ inj dim $((R/P)_P) \leq k - 1$. Therefore by induction,

K-dim(R) = K-dim(R_P) + 1 = gl dim(R_P) + 1 $\leq$ k = gl dim(R).

For equality of the two dimensions, we shall show $Ext^{k-1}_{R_P}((R/P)_P,(R/P)_P) \neq 0$. Again consider the exact sequence

$$0 \to R/P \overset{x}{\to} R/P \to R/Rx+P \to 0. \qquad (2)$$

Let S be a simple submodule of R/Rx+P. Applying the functor Ext(-,R/J) we get a long exact sequence

$$\ldots \to Ext^k_R(R/Rx+P,R/J) \to Ext^k_R(S,R/J) \to 0.$$

Since by [4] and prop. 3 $Ext^k_R(S,R/J) \neq 0$, $Ext^k_R(R/Rx+P,R/J) \neq 0$. Apply the functor Ext(-,R/J) to (2). Since $Ext^k_R(R/P,R/J) = 0$, $Ext^{k-1}_R(R/P,R/J) \neq 0$. By induction, it follows that $Ext^{k-1}_R(R/P,N) \neq 0$ for all modules N of finite length. Since $Ext^k_R(R/P,R/P) = 0$, from (2) we get $Ext^{k-1}_R(R/P,R/P) \neq 0$.

We claim $Ext^k_R(R/Rx+P,R/P) \neq 0$. Since R/Rx+P has a simple submodule, it suffices to prove $Ext^k_R(R/J,R/P) \neq 0$. By applying the functor Ext(R/J,-) to (2), we get

$$\ldots \to Ext^k_R(R/J,R/P) \to Ext^k_R(R/J,R/Rx+P) \to 0.$$

If follows from the fact that R/Rx+P has a simple factor module that $Ext^k_R(R/J,R/Rx+P) \neq 0$ and therefore $Ext^k_R(R/J,R/P) \neq 0$.

Let K be the factor module of $(R/P)_P$ derived from the exact sequence

$$0 \to R/P \to (R/P)_P \to K \to 0 \ (3).$$

Applying the functor Ext(R/Rx+P,-) to (3), we get

$$\cdots \to \mathrm{Ext}_R^{k-1}(R/Rx+P,(R/P)_P) \to \mathrm{Ext}_R^{k-1}(R/Rx+P,K) \to \mathrm{Ext}_R^{k}(R/Rx+P,R/P)$$
$$\to \mathrm{Ext}_R^{k}(R/Rx+P,(R/P)_P) \to \cdots .$$

Since R is a stable ring, no injective in the injective resolution of $(R/P)_P$ contains a simple module. Therefore $\mathrm{Ext}_R^k(R/Rx+P,(R/P)_P) = 0$ and $\mathrm{Ext}_R^{k-1}(R/Rx+P,K) \neq 0$.

It follows from applying the functor Ext(-,K) to (2) that $\mathrm{Ext}_R^{k-1}(R/P,K) \neq 0$. Finally apply the functor Ext(R/P,-) to (3). We get

$$\cdots \to \mathrm{Ext}_R^{k-1}(R/P,R/P) \to \mathrm{Ext}_R^{k-1}(R/P,(R/P)_P) \to \mathrm{Ext}_R^{k-1}(R/P,K) \to 0$$

Therefore, $\mathrm{Ext}_R^{k-1}(R/P,(R/P)_P) \cong \mathrm{Ext}_{R_P}^{k-1}((R/P)_P,(R/P)_P) \neq 0$ and we are done. □

Corollary 8: If R is a stable FBN ring with finite global dimension, then for each finitely generated left R-module M K-dim(M) ≤ inj dim(M).

Proof: Again, by [6, Prop. 1, Cor. 3] it suffices to assume R has a unique maximal ideal J such that R/J is a simple artinian ring. By theorem 7 and [4, Lemma 3], we have K-dim(M) ≤ K-dim(R) = gl dim(R) = inj dim(R/J).

Consider the exact sequence

$$0 \to JM \to M \to M/JM \to 0.$$

By applying the functor Ext(R/J,-), we get the long exact sequence

$$\cdots \to \mathrm{Ext}_R^k(R/J,M) \to \mathrm{Ext}_R^k(R/J,M/JM) \to 0$$

where $k = \text{inj dim}(R/J)$. Since $\text{Ext}_R^k(R/J,M/JM) \neq 0$ by proposition 3, $\text{Ext}_R^k(R/J,M) \neq 0$. Therefore $\text{K-dim}(M) \leq \text{inj dim}(M)$. □

3. Stable Rings with Finite Global Dimension: Suppose a noetherian ring R is classically local with unique maximal ideal J. We can complete R in the J-adic topology. Call this completion $\hat{R}$. Moreover, any left finitely generated module can be completed with respect to the filtration $\{J^n M\}$. Call this completion $\hat{M}$. It is an open question whether R noetherian implies $\hat{R}$ noetherian under these hypotheses. However, the following result of Lambek and Michler gives us a partial result.

Theorem 9: [19, Thm. 6] If R is a classically local noetherian ring with unique maximal ideal J, then the J-adic completion of R is ring isomorphic to the bicommutator of E(S) where S is the unique simple left R-module. □

Corollary 10: If R is a classically local noetherian ring with unique maximal ideal J and unique simple R-module S such that E(S) is artinian, then $\hat{R}$ is left noetherian. Furthermore, E(S) produces a Morita duality between the rings $\hat{R}$ and $T = \text{End}(E(S))$

Proof: By Müller [22, Thm. 8], E(S) produces a Morita duality. Therefore $E(S)_T$ is artinian [1, Thm. 24.5] and $\hat{R} = \text{Hom}_T(E(S),E(S))$ is left noetherian [1, Thm. 24.5]. □

Corollary 11: Every classically local noetherian PI ring has a right and left noetherian completion.

Proof: By Vamos and Jategaonkar, a classically local noetherian PI ring has artinian minimal injective cogenerator [30 or 15]. □

Since for a classically local noetherian ring R, the unique maximal ideal J satisfies the Artin-Rees property, we find that, for every submodule N of finitely generated left R-module M and for each $k \in \mathbb{N}$, there exists an $n \in \mathbb{N}$ such that

$$N \cap J^n M \subseteq J^k N$$

[21, Cor. 2.14]. Therefore, the results on completion in [2] still hold. In particular,

Theorem 12: If R is a classically local noetherian ring with unique maximal ideal J, then the following properties hold:

(a) $\hat{R}$ is flat as a (left) right R-module

(b) If M is finitely generated R-module, then $\hat{M} \cong \hat{R} \otimes_R M$.

(c) If $\hat{R}$ is left noetherian, then $\hat{R}$ is semiperfect left classically local with $\hat{J}$, its unique maximal ideal.

(d) $\hat{R}/\hat{J}^n \cong R/J^n$

Proof: The proofs of (a) and (b) mimic the proofs on completions found in [2]. Goldie [11, Thm. 4.6] has proved that $\hat{J}$ is the unique maximal ideal of $\hat{R}$ and $R/J^n \cong \hat{R}/\hat{J}^n$. If $\hat{R}$ is left noetherian, a result of Jategaonkar [16, Thm. 1.1] shows that $\hat{J}$ satisfies the left Artin-Rees property. Thus, in this case, $\hat{R}$ is left classically local. The proof that $\hat{R}$ is semiperfect, i.e. idempotents lift mod $\hat{J}$, can be found in [8, Prop. 21.7B].

The following proposition is the key to determining the primeness of a classically local noetherian ring with finite global dimension and the proof can be found in [5, Thm. 10.6].

Proposition 13: If R is a left classically local semiperfect noetherian ring with finite global dimension, then R has unique minimal prime ideal. □

We generalize this result to semiprime classically local noetherian rings with finite global dimension whose J-adic completion is noetherian.

Proposition 14: Let R be a semiprime classically local noetherian ring with gl dim(R) < ∞. If $\hat{R}$ is left noetherian, then R is a prime ring.

Proof: Suppose R satisfies the hypothesis of the theorem and let J be the unique maximal ideal. By [4, Lemma 3 and Corollary] inj dim(R/J) = proj dim(R/J) = gl dim(R). Let

$$0 \to P_n \to \cdots \to P_1 \to P_0 \to R/J \to 0$$

be a finitely generated projective resolution of ${}_R R/J$. Since $\hat{R}_R$ is flat (in fact, faithfully flat),

$$0 \to \hat{P}_n \to \cdots \to \hat{P}_1 \to (\widehat{R/J}) \to 0$$

is a projective resolution of $(\widehat{R/J}) \cong \hat{R}/\hat{J}$. Therefore proj dim(R/J) < ∞. Again by [4, Corollary] and theorem 12, proj dim($\hat{R}/\hat{J}$) = gl dim($\hat{R}$). Therefore, $\hat{R}$ is a semiperfect, left classically local, left noetherian ring with finite global dimension. By proposition 13, $\hat{R}$ has a unique minimal prime ideal N. Therefore, N is nilpotent and $N \cap R = 0$. Suppose $xRy = 0$. By the construction of $\hat{R}$, $x\hat{R}y = 0$. Since N is a prime ideal, $x \in N$ or $y \in N$, and hence Rx or Ry is nilpotent. Since R is semiprime, this implies $x = 0$ or $y = 0$. □

Corollary 15: Let R be a semiprime classically local noetherian ring with finite global dimension. If E(S) is artinian where S is the unique simple left R-module, then R is a prime ring.

Proof: Follows from proposition 14 and corollary 10. □

Corollary 16: If R is a semiprime classically local noetherian PI ring with finite global dimension, then R is a prime ring.

Proof: Corollary 11 and Proposition 14. □

Corollary 17: Let R be a semiprime classically local noetherian ring with finite global dimension. If the unique maximal ideal J has a normalizing set of generators, then R is a prime ring.

Proof: By [16,Thm. 1.4], in such a case $\hat{R}$ is left noetherian. The result now follows from proposition 14. □

For stable rings, we get sharper results as the next series of propositions will show.

Proposition 18: If R is a left artinian, left stable ring with finite left global dimension, then R is a semisimple ring.

Proof: Suppose R is left stable left artinian ring with finite left global dimension. By [6, Cor. 3, Cor. 6, and Cor. 4], it suffices to consider the global dimension of each of the localizations of R. Suppose R is a classically local left artinian ring. By [10, Prop. 17.8], R has finite left global dimension. Let S be a simple left submodule of R. Therefore proj dim(R/S) = proj dim(S) + 1 or S is projective. Since by [4, Corollary] the former gives a contradiction, gl dim(R) = 0.

Proposition 19: Let R be a stable noetherian ring. If R has finite global dimension, then R is a semiprime ring.

Proof: Let $0 \neq x \in R$ such that the left annihilator of Rx is a prime ideal P. Since R is a stable noetherian ring, each ideal has the left and right Artin-Rees property. By [27, Prop 3.3], we have a perfect left localization at each prime ideal P. Since P_P satisfies the left Artin-Rees property in R_P [20, Theorem 5], the left global dimension of R_P equals the left projective dimension of R_P/P_P [4, Corollary]. But $(Rx)_P$ is a non-trivial semisimple module of the left noetherian ring R_P. Hence, ℓ.gl dim(R_P) = proj dim($(Rx)_P$). Thus either proj dim($R_P/(Rx)_P$) = 1 + proj dim($(Rx)_P$) or $(Rx)_P$ is a projective module. Since the former would give a contradiction, $(Rx)_P$ is a projective module and therefore R_P is a semisimple ring. Thus P is a minimal prime ideal of R. By [28, Thm. 2], R has an artinian classical ring of quotients Q(R). By [10, Prop. 17.8], Q(R) has finite global dimension and an argument similar to that given in proposition 1, shows Q(R) is left stable. By proposition 18, Q(R) is a semisimple ring. Therefore by Goldie's theorem [29, Thm. II. 2.2], R is semiprime. □

Corollary 20: Let R be a stable classically local noetherian ring with finite global dimension. If $\hat{R}$ is left noetherian, then R is a prime ring. □

We remark that proposition 19 is true in a more general case. If every prime ideal of R satisfies the Artin-Rees property, then the proof of proposition 19 still holds. Of course, for FBN rings, this property is equivalent to stability.

We conclude with a structure theorem for stable noetherian PI rings with finite global dimension.

Theorem 21: If R is a stable noetherian PI ring with finite global dimension, then R decomposes into a finite direct product of prime rings.

Proof: Since R is semiprime, R has a finite set of minimal prime ideals. Let $P_1,\dots,P_n$ be the set of minimal prime ideals of R. Let M be a maximal ideal of R. By [10, Prop. 17.8], and proposition 1, R_M is a stable noetherian PI ring with finite global dimension. By corollary 20 and corollary 16, R_M is a prime ring. Therefore, each maximal ideal can contain only one minimal prime ideal. By proposition 19 and the Chinese Remainder Theorem [8, Thm. 18.30], $R = R/\cap P_i \cong \pi\ R/P_i$. □

We conjecture that theorem 21 is true for all noetherian stable rings with finite global dimension. By proposition 14, it would suffice to show that the completions of the localizations are noetherian. We conjecture this also to be true.

Remark: It has been pointed out to the authors that theorem 21 can be proved for all noetherian stable rings with finite global dimension without showing that completions of localizations are noetherian. A theorem by Ramras [5, Corollary 10.7] in conjunction with proposition 14 solves the problem for the general case.

References

[1] F. W. Anderson and K. R. Fuller, Rings and Categories of Modules, Graduate Texts in Mathematics, Springer-Verlag, Berlin-Heidelberg-New York, 1974.

[2] M. F. Atiyah and I. G. MacDonald, Introduction to Commutative Algebra, Addison-Wesley, 1969.

[3] H. Bass, On the ubiquity of Gorenstein rings, Math. Z., 82 (1963), 8-28.

[4] M. Boratynski, A change of rings theorem and the Artin-Rees property, Proc. Amer. Math. Soc., 53 (1975), 307-310.

[5] A. W. Chatters and C. R. Hajarnavis, Rings with Chain Conditions, Pitman, Boston-London-Melbourne, 1980.

[6] R. Damiano and Z. Papp, On consequences of stability, Comm. Algebra, 9 (1981), 747-764.

[7] D. Z. Djokovic, Epimorphisms of modules which must be isomorphisms, Canad. Math. Bull., 16 (1973), 513-515.

[8] C. Faith, Algebra II: Ring Theory, Springer-Verlag, Berlin-Heidelberg-New York, 1976.

[9] P. Gabriel, Des Categories abeliennes, Bull. Soc. Math. France, 90 (1962), 323-448.

[10] J. Golan, Localization of Noncommutative Rings, Marcel Dekker, New York, 1975.

[11] A. W. Goldie, Localization in non-commutative noetherian rings, J. Alg. 5 (1967), 89-105.

[12] R. Gordon and J. C. Robson, Krull dimension, Amer. Math. Soc. Memoirs, #133 (1973).

[13] A. V. Jategaonkar, Injective modules and localization in noncommutative noetherian rings, Trans. Amer. Math. Soc. 190 (1974), 109-123.

[14] A. V. Jategaonkar, Relative Krull dimension and prime ideals in right noetherian rings, Comm. Alg., 2 (1974), 429-468.

[15] A. W. Jategaonkar, Certain injectives are artinian, Noncommutative ring theory, Lecture Notes in Math. No. 545, 128-139.

[16] A. V. Jategaonkar, Morita duality and noetherian rings, J. Alg. 69 (1981), 358-371.

[17] S. Jöndrup, Homological dimensions of some P.I. rings, Comm. Algebra, 8 (1980), 685-696.

[18] G. Krause, On fully left bounded left noetherian rings, J. Algebra, 23 (1972), 88-99.

[19] G. Krause, T. H. Lenagan, and J. T. Stafford, Ideal invariance and artinian quotient rings, J. Algebra, 55 (1978), 145-154.

[20] J. Lambek and G. Michler, Completions and classical localizations of right noetherian rings, Pac. J. Math. 48 (1973), 133-140.

[21] J. C. McConnell, The noetherian property in complete rings and modules, J. Alg. 12 (1969), 143-153.

[22] B. Müller, On Morita duality, Canad. J. Math., 21 (1969), 1338-1347.

[23] B. Müller, Ideal invariance and localization, Comm. Algebra, 7 (1979), 415-441.

[24] B. Müller, Two-sided localization in noetherian PI rings, J. Algebra 63 (1980), 359-373.

[25] R. Resco, L. Small and J. T. Stafford, Krull and global dimensions of semiprime noetherian PI-rings, preprint.

[26] J. J. Rotman, An Introduction to Homological Algebra, Academic Press, New York, 1980.

[27] P. F. Smith, Localization and the AR property, Proc. London Math. Soc., 22 (1971), 39-68.

[28] P. F. Smith, On two-sided artinian quotient rings, Glasgow Math. J., 13 (1972), 159-163.

[29] B. Stenström, Rings of Quotients, Springer-Verlag, Berlin-Heidelberg-New York, 1975.

[30] P. Vamos, Semi-local noetherian PI-rings, Bull. London Math. Soc., 9 (1977), 251-256.

[31] R. Walker, Local rings and normalizing sets of elements, Proc. London Math. Soc., 24 (1972), 27-45.

SUMS OF UNIFORM MODULES

John Dauns
Department of Mathematics
Tulane University
New Orleans, LA 70118/USA

INTRODUCTION. The threefold objective of this note is not only (1) to give some new recent developments, but also (2) to acquaint the reader with an already existing theory, and then (3) suggest further new problems and unanswered questions. This mostly expository article originated from two sources both of which study very similar classes of modules. First, Section 2 here suggests that the already well developed theory of mixed abelian groups ([14; part II]) might possibly have some analogues for modules M over more general rings R. The analogue of the torsion subgroup now is the singular submodule ZM. Maximal torsion free submodules $C, D \subset M$ are investigated, where M/C is essentially torsion (i.e. $Z(M/C) \subset M/C$ is essential). Those readers, who object to the unavoidable fact that C and D might not be isomorphic, should work in the so-called category Warf (as in [14; part II]) whose objects are the usual modules. However, a morphism of C into D is a pair A,α where $A \subset C$ is a submodule such that $Z(C/A) = C/A$, while $\alpha : A \longrightarrow D$ is an ordinary module map. In Warf, C and D become isomorphic objects.

Secondly, in Section 3, this article generalizes the concept of the socle of a module by replacing simples with torsion free uniform modules. These general results are applied in the next Section 4 to the particular special case when the module M is the ring R regarded as a module over itself. As a further application, properties of the endomorphism ring of a module is related to its structure.

One reason why this particular topic was chosen is that there seem to be some open questions here which possibly would be answered. For example, could Section 2 be used as a spring-board to generalize [14; part II] to modules? If for any module M whatever, the submodule $ZM \subset Z_2M \subset M$ is defined by $Z_2M/ZM = Z(M/ZM)$, then $Z(M/Z_2M) = 0$. Is enough gained by replacing ZM with Z_2M in order to justify the added technical complexity?

It is shown at the end of Section 4 that appropriate primeness hypotheses on M will guarantee that the injective hulls of all uniform submodules of M will be isomorphic. So far little has been done to investigate how various specialized kinds of uniform modules (as in [17] [20], [6], and [9]) relate to various primeness concepts (such as in

[11], [22], [24], [4], [5], [7], and [8]).

The type of modules and rings considered here (in 3.1) are subdirect products of indecomposable injective modules. There are various specialized types of subdirect products ([2; p.258], [12], [16; p.115], and [22; p.65]). Under what additional hypotheses if any will the modules and rings discussed here be one of these special types of subdirect products?

There are several additional areas the reader may wish to explore further. The categories invented in [27] allow one to treat modules M of the form $M = E(\oplus_I U)$ as if they were simply direct sums. In another direction, rather than defining a module M to be torsion if $ZM = M$, or as already suggested if $Z_2M = M$, one could use more general torsion theories to define torsion modules like in [25]. Do results analogous to those satisfied by the injective hull as given here also hold for the quasi-injective hull (see [10], [19], and [20])?

1. A CLASS OF MODULES

Notation and terminology is established in which questions can be easily raised and concisely answered about a certain class of modules. Successively more hypotheses are put on the module as this exposition goes on, starting with the general and ending with the more specialized. Fundamental lemmas are proved with as few assumptions as possible, even if later they are used under more restrictive hypotheses.

1.1 NOTATION. By a module will be meant a right unital module over the ring R, and a left module over its R-endomorphism ring $\mathrm{End}\, M_R = \mathrm{End}\, M$. The symbol $<$ denotes right R-submodules $N < M$. If $M = R$, then $<$ denotes right ideals; whereas " $\lhd$ " is reserved for ideals. Large, or essential, right R-submodules are denoted by a symbol similar to the one used for submodules, except larger-- "$<<$". The notation "$A \not<< B$" means that $A < B$, but that A is <u>not</u> large in B. Here "$<$", "$\lhd$", "$<<$", and "$\subset$" is used in the Bourbaki sense, where equality is also allowed.

For $m \in M$ and $N < M$, $m^{-1}N < R$ is defined as the inverse image $m^{-1}N = \{r \in R \mid mr \in N\} < R$ of N under the R-map $R \longrightarrow M$, $r \longrightarrow mr$. By now, the notation $m^{\perp} \equiv \{r \in R \mid mr = 0\} < R$ is customary, although, also $m^{\perp} = m^{-1}(\{0\}) \equiv m^{-1}0$. Hence, for $m + N \in M/N$, $(m + N)^{\perp} = m^{-1}N$.

The injective hull with respect to R of a right R-module M is denoted by both $\hat{M}$ and $E(M) = EM$, the latter notation being used when M is a complicated quotient module or direct sum. The usual singular submodule $ZM = \{m \in M \mid m^{\perp} << R\}$ will also be called the <u>torsion submodule</u> of M, even though "Z" does not have a hereditary torsion theory. Thus in this terminology, a module B is torsion if $ZB = B$ and torsion free if $ZB = 0$. The operations "<", "<<", "$^{-1}$", "$^{\perp}$", "E", "^", and "Z" apply to right R-modules and always in the ring R, and never in other rings, like $\hat{R}$ etc.

1.2 SUMS OF UNIFORM SUBMODULES. Consider a given fixed module M and sets I, J of torsion free uniform submodules of M whose sum is a direct sum, i.e., if H is defined as $H \equiv \sum\{U \mid U \in I\}$, then $H = \oplus\{U \mid U \in I\}$. Furthermore, take the set I to be maximal with this property. In other words, $I = \{U < M \mid U$ is uniform, $ZU = 0\}$ such that (i) $\sum\{I < M \mid U \in I\} = \oplus\{U \mid U \in I\} \equiv H$; and (ii) for any uniform $V < M$ with $ZV = 0$, necessarily $H \cap V \neq 0$. For simplicity abbreviations of the type $\oplus\{U \mid U \in I\} = \oplus U$ will be used. Sums of the type $\oplus U$ will be over the biggest possible index set; here over all of I, and never over a proper subset of I.

For $U < M$, the smallest complement submodule $\bar{U} < M$ containing U is simply any maximal essential extension in M containing $U << \bar{U} < M$. The set $\bar{I} = \{\bar{U} \mid U \in I\}$ is also maximal in the above sense with respect to (i) and (ii). Even though $H \subset \oplus\{\bar{U} \mid \bar{U} \in \bar{I}\} \equiv \tilde{H}$ both sets I and $\bar{I}$ are maximal and in general incomparable, i.e. $I \not\subset \bar{I}$. Now, however, $\tilde{H}$ is a maximal or biggest direct sum of torsion free uniform submodules of M. Since $H << \tilde{H}$, it turns out that what is important is that only the I be maximal. The sum H derived from I need not be maximal. Although neither I nor H are unique, their significance is that what will be said will be true for every such choice of I. Furthermore, they will be useful tools for uncovering the properties of the type of modules described below.

1.3 ASSUMPTIONS. From now on it will be assumed that every nonzero torsion free submodule of M contains a uniform submodule.

2. TORSION AND COMPLEMENT SUBMODULES

The notation of 1.2 for I and $H = \oplus U$ and the property 1.3 for M will be continued throughout. Since M usually does not have a unique torsion free part, one is forced to use maximal torsion free submodules C of M. These are characterized, and their properties described, as well as those of their quotients M/C. What makes complement submodules $K < M$ useful not only here but also later in some of the proofs in Section 4 is that if $B < M$ with $B \oplus K << M$, then the image of B remains essential upon passage to quotients $(B \oplus K)/K << M/K$. (For definitions and properties of complements, see [9; 1.1 and 1.2].)

The special case of the next lemma when $ZM \subset A$ is easy to prove. The next two lemmas may be of independent interest because they apply to any module M without invoking any restrictive hypotheses on M, such as 1.3.

2.1 LEMMA. Assume that $A << B < M$ and define $A, B < M$ as $Z(M/A) = \overline{A}/A$ and $Z(M/B) = \overline{B}/B$. Then $\overline{A} = \overline{B}$, and

$$\frac{Z(M/A)}{B/A} \cong Z(\frac{M}{B}) .$$

2.2 LEMMA. For any module M and for I and H as in 1.2, suppose that $C < M$ is any submodule of M with $ZC = 0$. Then the following are all equivalent.

(i) $\forall$ torsion free uniform $V < M$, $\Longrightarrow V \cap C \neq 0$.

(ii) $\forall U \in I$, $\Longrightarrow U \cap C \neq 0$

(iii) $H \cap C << H$.

2.3 COROLLARY. If in addition, M also satisfies 1.3, then (iv) below is also equivalent to 2.2(i)-(iii) above:

(iv) $\forall$ torsion free $W < M \Longrightarrow W \cap C \neq 0$.

(v) Furthermore, $H \cap C << C$.

By the last lemma and corollary, in the next theorem condition (4) below can be supplemented by three additional equivalent conditions.

2.4 THEOREM I. If the module M has the property 1.3 that every torsion free submodule contains a uniform submodule, then the following four conditions (1)-(4) are equivalent.

(1) $C < M$ is a maximal torsion free submodule of M.

(2) $C < M$ is maximal such that $ZC = 0$, but $Z(M/C) << M/C$.

(3) $C < M$ is a torsion free complement such that M/C contains no torsion free uniform submodules.

(4) $C < M$ is a torsion free complement submodule which intersects every torsion free uniform submodule of M nontrivially.

Since 1.3 is not assumed in the next corollary it may be of independent interest.

2.5 COROLLARY 1 TO THEOREM I. For any unital module M and any submodule $C < M$, the above conditions 2.4(1), (2) and (3) are all equivalent.

2.6 COROLLARY 2 TO THEOREM I. If M is as in 1.3 and $C < M$ satisfies one and hence all the conditions 2.4 (1)-(4), then there exists a set $J = \{V\}$ of torsion free uniform submodules $V < M$ which is maximal (as in 1.2) with respect to its sum being a direct sum, and furthermore such that $\oplus\{V \mid V \in J\} \subset C$.

2.7 COROLLARY 3 TO THEOREM I. Every module M having the property 1.3 contains an intrinsic submodule $A < M$ which depends only on M such that for every submodule $C < M$ as in the last thoerem satisfying 2.4 (1)-(4), as well as for every choice of maximal set $I = \{U\}$ as in 1.2, the following hold.

(i) $C \oplus ZM \subset A$;

(ii) $Z(M/C) = A/C$;

(iii) $Z(M/\oplus\{U \mid U \in I\}) = A/\oplus\{U \mid U \in I\}$.

3. INTRINSIC SUBMODULES

Just for the moment, suppose that M satisfies 1.3 but is not torsion free, i.e., $ZM \neq 0$. 1) Then the sum of two torsion free uniform submodules of M need not be torsion free. 2) If $B < M$ with $ZB = 0$, then $\hat{M}$ may contain two distinct injective hulls $\hat{B}$, $EB \subset M$ of B with $B \subset \hat{B} \cap EB$. 3) There may exist three torsion free uniform submodules $U,V,W \subset M$ such that $W \subset U + V$, $\hat{U} \cong \hat{V}$, but $\hat{W} \ncong \hat{U}$. The above three pathologies will be eliminated by assuming in addition to 1.3, that M is also torsion free throughout from now on to the end of this exposition.

Although the direct sum $H = \oplus\{U \mid U \in I\} < M$ depends on one's particular choice of I, it will be shown that the submodule $\oplus\{\hat{U} \mid U \in I\} < \hat{M}$ is completely independent of the choice of I. Moreover, it will be characterized abstractly and it will be seen to possess some interesting properties.

From now on all the way to the end the following standard standing hypothesis will be automatically assumed for M.

3.1 STANDARD HYPOTHESIS. Assume that (i) $ZM = 0$, and that (ii) every nonzero submodule of M contains a uniform submodule.

The above two assumptions are equivalent and may be replaced by the single hypothesis that M contains an essential direct sum of torsion free uniform submodules.

The previous notation and definition for I and H as in 1.2 will be continued. If C is a maximal torsion free submodule of a module which only satisfies our previous weaker assumption 1.3, then C satisfies the stronger hypothesis 3.1.

3.2. For any modules A and B whatever, A and B will be said to be <u>parallel</u> -- written as $A \| B$ -- provided that A does not contain a nonzero submodule isomorphic to some submodule of B.

The next two facts will be needed in order to understand the definition of some intrinsic submodules of the module M. (See [9; 2.3 and 2.5] for proofs.)

3.3. Suppose that A, B, C, and D are any modules with $A < D$, $B < D$; with $A \| C$ and $B \| C$ where $ZC = 0$. Then likewise also $(A + B) \| C$.

3.4. Let $\{A_\alpha\}$ be a set of submodules of some module C and let $\{B_\beta\}$ be another set of submodules of some other fixed module. Assume that for all α and β

$$ZB_\beta = 0, \; A_\alpha \| B, \quad \text{and} \quad ZC = 0.$$

Then

$$\textstyle\sum_\alpha A_\alpha \| \sum_\beta B_\beta \; .$$

3.5 CLASSES OF UNIFORM MODULES. Two modules U and V will be said to be related--written as $U \sim V$ -- if i) U and V are uniform and ii) if $\hat{U} \cong \hat{V}$. Let $\Omega = \{\alpha, \beta, \ldots\}$ denote the equivalence classes modulo the equivalence relation "$\sim$" of related uniform submodules of the fixed torsion free module M. Every equivalence class $\alpha \in \Omega$ defines an intrinsic submodule $M_\alpha = \sum\{U \mid U \in \alpha\} < M$. Let $\hat{\Omega}$ be the analogue of Ω for the module $\hat{M}$. Since $\alpha \longrightarrow \hat{\alpha} \equiv \{V < \hat{M} \mid \hat{V} \cong \hat{U}, U \in \alpha\}$ is a bijection of $\Omega \longrightarrow \hat{\Omega}$, the same index set $\Omega = \{\alpha, \beta, \ldots\}$ will be used for M, $\hat{M}$, and for various submodules of $\hat{M}$.

Every torsion free module M contains another type of intrinsic submodule $U(M) = UM < M$ defined as $UM = \sum\{U \mid U < M$ is uniform$\}$. Thus also $UM = \sum\{M_\alpha \mid \alpha \in \Omega\}$.

The next fact and lemma will be used several times later on.

3.6. For any modules $A < B$ with $ZB = 0$, there is a unique injective hull $\hat{A}$ of A such that $A \subset \hat{A} \subset \hat{B}$.

3.7 LEMMA. For any maximal set I as in 1.2 and M as in 3.1, suppose that $V \subset \hat{M}$ is any uniform submodule with $V \in \beta \in \Omega$. Then

(i) for any $U_1, \ldots, U_m \in I$

$$\hat{V} \subset \hat{U}_1 \oplus \ldots \oplus \hat{U}_m \Longleftrightarrow V \cap (U_1 \oplus \ldots \oplus U_m) \neq 0.$$

(ii) There exists a unique minimal subset $\{U_1, \ldots, U_n\} \subset I$ such that $\hat{V} \subset \hat{U}_1 \oplus \ldots \oplus \hat{U}_n$, in which case $U_1, \ldots, U_n \in \beta$.

The next lemma mildly generalizes [15; p.207, Lemma 3.3] and can be proved similarly, or directly.

3.8 LEMMA. If U is any uniform module, and W any other module, let $X = \{\phi \in \mathrm{Hom}_R(U,W) \mid U \cap \phi^{-1}0 \neq 0\}$, define XU to be the submodule $XU = \sum\{\phi U \mid \phi \in X\} < W$. If $\lambda : U \longrightarrow W$ is an arbitrary homomorphism, then

(i) $XU \subset ZW$; in particular, if $ZW = 0$, then

(ii) a) EITHER $\lambda \equiv 0$

b) OR λ is monic and $U \cong \lambda U$.

3.9 COROLLARY. If R is a ring with $ZR = 0$ and $U < R$ is a uniform right ideal, then for any $x \in R$, either $xU = 0$, or $U \longrightarrow xU$, $u \longrightarrow ux$ is an isomorphim and $U \cong xU$.

3.10 COROLLARY. On the class of all torsion free modules, U as in 3.5 is a functor.

PROOF. If $ZA = 0$, $ZB = 0$, and $f : A \longrightarrow B$ is a module homomorphism, then by 3.8 (ii)b), $f(UA) \subset UB$. Thus let $\cdot Uf : UA \longrightarrow UB$ be the restriction and corestriction.

Next, some general concepts are clarified which later will relate to the submodules $M_\alpha < M$ and $(EM)_\alpha < EM$.

3.11 FACTS.

(1) For any module W, if $\tilde{W}$ denotes its quasi-injective hull as a submodule of $\hat{W}$, then $\tilde{W}$ is the unique submodule $\tilde{W} = (\text{End } \hat{W})W$. Thus $\tilde{W}$ is the unique smallest submodule of $\hat{W}$ containing W which is invariant under every R-endomorphism from End $\hat{W}$.

(2) If $\{W_\alpha\}$ is any indexed set of modules whatever such that $\text{Hom}_R(W_\alpha, W_\beta) = 0$ for every pair $\alpha \neq \beta$, then

$$(\oplus W_\alpha)^\sim = \oplus \tilde{W}_\alpha .$$

3.12 THEOREM II. With the above notation and for M as in 3.1, the following hold for all $\alpha \in \Omega$:

(i) $UM = \oplus\{M_\alpha \mid \alpha \in \Omega\}$;

(ii) $\oplus M_\alpha << M$;

(iii) $M_\alpha < M$ is fully invariant;

(iv) $\text{Hom}_R(M_\alpha, M_\beta) = 0$ for $\alpha \neq \beta \in \Omega$;

(v) $(\oplus M_\alpha)^\sim = \oplus \tilde{M}_\alpha$;

(vi) $(UM)_\alpha = M_\alpha$; $(EM)_\alpha \subset E(M_\alpha)$.

PROOF. (i) and (ii) are proved in [9; 2.9]. Conclusion (iii) is a consequence of 3.8 (ii). Use of 3.4 gives (iv), which when combined with 3.11 (2) gives (v).

(vi) By definition of M_α, $M_\alpha = (UM)_\alpha$. If $W \in \alpha$ and $Y \in \hat{\alpha} = \{V < \hat{M} \mid \hat{V} \cong \hat{W}\}$, then $Y \cap M \in \alpha$, and thus $0 \neq Y \cap M \subset M_\alpha$. By 2.1 there are unique injective hulls in $\hat{M}$ and $Y = E(Y \cap M) \subset E(M_\alpha)$. Since by definition $(EM)_\alpha = \sum\{Y \mid Y \in \hat{\alpha}\}$, it now follows that $(EM)_\alpha \subset E(M_\alpha)$.

The next proposition gives very detailed direct sum decompositions for the two types $U(EM)$ and $(EM)_\alpha$ of intrinsic submodules of the injective hull EM of our module M. Conclusion 3.13 (ii) is given in [17; 5.10]; its proof must depend heavily on $ZM = 0$.

3.13 PROPOSITION. Suppose that M is any module which contains an essential direct sum of torsion free uniform submodules. Let I, $\oplus\{U \mid U \in I\} << M$, and Ω be as in 1.2 and 3.5; let U be the functor in 3.5 and 3.10. Then the following hold for all $\alpha \in \Omega$ and every maximal set I.

(i) $(EM)_\alpha = \oplus\{\hat{U} \mid U \in \alpha \cap I\}$.

(ii) $U(EM) = \oplus(EM)_\alpha = \oplus\{\hat{U} \mid U \in I\}$.

(iii) $U(EM)$ and all the $(EM)_\alpha$ are fully invariant submodules of EM, and hence quasi-injective.

(iv) $\mathrm{Hom}_R((EM)_\alpha, (EM)_\beta) = 0$ for $\alpha \neq \beta \in \Omega$.

PROOF. Only (ii) will be proved modulo the previously stated facts and lemmas. (ii). Replacement in 3.12 (i) of M with EM shows that $UEM = \oplus\{(EM)_\alpha \mid \alpha \in \Omega\}$, where $(EM)_\alpha = (UEM)_\alpha$. Define a module $G \subset EM$ as $G = \oplus\{\hat{U} \mid U \in I\}$. Hence $G \subset UEM$ already by definition of G. Thus, in order to show that $G = UEM$, it now simply suffices to show that G contains every uniform submodule of EM. But if $V < EM$ is uniform, then by 3.7 (ii), $\hat{V} \subset \hat{U}_1 \oplus \ldots \oplus \hat{U}_n \subset G$ for some subset $\{U_1, \ldots, U_n\} \subset I$. Thus $G = UEM$.

4. DIRECT SUMS AND RINGS

The previous results are applied to the ring R here regarded as a right module over itself. A class of rings R more general than the right Noetherian is considered. The following question is raised and answered. Are there (necessarily proper) natural classes of R-modules such that for every member M of this class, the injective hull of M is a direct sum of indecomposable injectives?

4.1 TERMINOLOGY. A right R-module W has uniform dimension n (also called Goldie dimension) -- denoted by $ud(W) \equiv udW = n$ -- if W contains an essential direct sum of n uniform submodules. By definition, a right ideal $L < R$ is prime if for any right ideals $A, B < R$ with $AB \subset L$, then either $A \subset L$ or $B \subset L$.

4.2 LEMMA. Suppose that U is a uniform torsion free module and that $ud(R/ZR) = m < \infty$. Then

(i) U is isomorphic to a submodule of $E(R/ZR)$. Hence

(ii) there are m or less distinct nonisomorphic indecomposable injective modules, i.e. $|\Omega| \leq m$.

The next ring theoretic proposition and the next theorem which follows from it are not tied down to the hypotheses of this article, and thus they might be of independent interest. Since it is possible to give short and very self-contained proofs, they will be both proved in complete detail.

4.3 PROPOSITION. If (a) $udR = m < \infty$, and $\{A_j \mid j \in J\}$ is any indexed set of right ideals which under set inclusion form an (b) upper directed partially ordered set (i.e. $i, j \in J \Longrightarrow k \in J$ such that $A_i + A_j \subset A_k$) with (c) $\cup\{A_j \mid j \in J\} << R$, then for some $n \in J$, already $A_n << R$.

PROOF. For every $j \in J$, $udA_j \le udR = m$. Choose an $i \in J$ such that udA_i is maximal. If $udA_i < m$, then $A_i \cap xR = 0$ for some cyclic uniform right ideal $xR < R$. By hypothesis (c), there exists an $r \in R$ with $0 \neq xr \in \cup\{A_j \mid j \in J\}$. Hence $0 \neq xr \in A_j$ for some $j \in J$. Choose a $k \in J$ with $A_i + A_j \subset A_k$. Thus $A_i \oplus xrR \subset A_k$. But then

$$udA_i < ud(A_i \oplus xrR) \le udA_k$$

violates the maximality of udA_i. Therefore $udA_i = udR = m$. The latter implies that $A_i << R$ as required.

A necessary and sufficient condition that a ring be right Noetherian is that every injective right module is a direct sum of indecomposable injectives. Since the class of rings R with $udR < \infty$ is noticeably more general than the Noetherian ones, every injective module over such rings R cannot be a direct sum of indecomposables. Consequently the class of all those R-modules whose injective hulls are direct sums of indecomposables is properly smaller than the class of all R-modules, if $udR < \infty$.

In view of the above, the next theorem and particularly its first corollary may be regarded as an attempt to generalize the structure theory of injective modules over Noetherian rings to a wider class of rings.

4.4. THEOREM III. Suppose that N is any torsion free, i.e. $ZN = 0$, right module over a ring R of finite uniform dimension $ud(R) < \infty$. If $F = \{W\}$ is any family of submodules $W < M$ whatever whose sum is an essential direct sum $\oplus\{W \mid W \in F\} << N$, then the injective hull $\hat{N}$ is obtained componentwise as $\hat{N} = \oplus\{\hat{W} \mid W \in F\}$.

PROOF. Since $\oplus\hat{W} << \hat{N} \subset \Pi\hat{W} \equiv \Pi\{\hat{W} \mid W \in F\}$, it would follow that $\hat{N} = \oplus\hat{W}$ if the latter were injective, which is the case if and only if $\oplus W$ has no proper essential extension in $\Pi\hat{W}$. If $\oplus\hat{W}$ is not injective, then there exists an $x = (x_f) = \{x_f \mid f \in F\} \in \Pi\hat{W} \setminus \oplus\hat{W}$ such that $\oplus\hat{W} << \oplus\hat{W} + xR$ is an essential extension. The latter implies that $x^{-1}(\oplus\hat{W}) << R$. (In fact, since $ZN = 0$, any extension of the form

$\oplus\hat{W} \subset \oplus\hat{W} + xR$ is essential if and only if $x^{-1}(\oplus\hat{W}) << R$ (see [7; p.165, 3.3]).) Let $J = \{i,j,k,\ldots\}$ denote the set of all finite subsets of F. For every finite subset $j \subset F$, define $A_j = \cap\{x_f^{\perp} \mid f \notin j\} < R$. (Thus $xA_j \subset \oplus\{W \mid W \in j\}$.) If $j \subset k$, then $A_j \subset A_k$. Thus not only under set inclusion is (b) $\{A_j \mid j \in J\}$ an upper directed partially ordered set of right ideals of R, but also the index set J again under set inclusion is isomorphic to it as a partially ordered set.

Furthermore, (c) $x^{-1}(\oplus\hat{W}) = \cup\{A_j \mid j \in J\} << R$. Since hypotheses (a), (b) and (c) of 4.3 hold, it now follows that for some finite subset $n \subset F$, $A_n << R$. But for all $f \in F$, if $f \notin n$, $A_n = \cap\{x_f^{\perp} \mid f \notin n\} \subset x_f^{\perp}$. Hence for every $f \notin n$, also $x_f^{\perp} << R$. Now, since $x \notin \oplus\hat{W}$, there exists an $f \notin n$ with $x_f \neq 0$. Thus $0 \neq x_f \in ZN = 0$ is a contradiction. Therefore $\hat{N} = \oplus\hat{W}$.

In the next corollary, without the additional uniform dimension hypothesis, the equalities (iii) and (iv) would have to be weakened to only $(EM)_\alpha \subset E(M_\alpha)$ and $U(EM) \subset E(UM)$ which is the case in Theorem II.

4.5 COROLLARY 1 TO THEOREM III. Suppose that M is a module containing an essential direct sum $\oplus\{U \mid U \in I\} << M$ of torsion free uniform submodules $U < M$ over a ring R with uniform dimension $udR = m < \infty$. If $\alpha \in \Omega$ and U are as before (in 3.5 and 3.10), then the following hold.

(i) $|\Omega| \leq m$; write $n = |\Omega|$, $\Omega = \{\alpha(1),\ldots,\alpha(n)\}$, and $M_j \equiv M_{\alpha(j)}$.

(ii) $\hat{M}_j = \oplus\{\hat{U} \mid U \in I \cap \alpha(j)\}$.

$\hat{M} = \oplus\{\hat{U} \mid U \in I\} = \hat{M}_1 \oplus \ldots \oplus \hat{M}_n$.

(iii) $(EM)_\alpha = E(M_\alpha)$.

(iv) $\hat{M} = U(EM) = E(UM)$.

The following facts, proved in [20; 36, 3.7, and 3.9] would be needed to deduce the next two corollaries in which M is specialized to $M = R$, as well as for understanding the structure of the ring $\hat{R}$.

4.6 FACTS. A ring R is called <u>regular</u> if for any $a \in R$, there is an $x \in R$ with $axa = a$. Below R is any ring having $ZR = 0$ and containing $1 \in R$, while $U < R$ will represent any uniform right ideal, if R has one at all. Then the following four facts hold.

(1) $\hat{R}$ is a regular ring.

(2) Over a regular ring R every uniform right ideal $U < R$ is a minimal right ideal.

(3) $\hat{U} = U\hat{R}$; hence $\hat{U}$ is also a right $\hat{R}$-ideal.

(4) The above (1)-(3) imply that $\hat{U}$ is a minimal right $\hat{R}$-ideal in $\hat{R}$.

In the next two corollaries the module M of both Theorems II and III is specialized to $M = R$.

4.7 COROLLARY TO THEOREM II. Suppose that $ZR = 0$, and that every right ideal of R contains a uniform right ideal. Let $\alpha, \beta \in \Omega$, $\alpha \neq \beta$, and let $\hat{R}_\alpha$ be defined as $\hat{R}_\alpha \equiv (ER)_\alpha$ (and not as $E(R_\alpha)$). Then the following hold

(i) $R_\alpha \lhd R$ and $\oplus R_\alpha \lhd R$.

(ii) $\hat{R}$ is a ring with $\hat{R}_\alpha \lhd \hat{R}$ and $\oplus\hat{R}_\alpha \lhd \hat{R}$.

(iii) a) $R_\alpha \subset R$ is a fully invariant right R-submodule of R with $R_\alpha R_\beta = 0$.

b) $\hat{R}_\alpha \subset \hat{R}$ is a fully invariant right $\hat{R}$-submodule of $\hat{R}$ with $\hat{R}_\alpha \hat{R}_\beta = 0$.

(iv) $\hat{U} = U\hat{R}$ is a minimal right $\hat{R}$-ideal of $\hat{R}$ for all $U \in I$.

Each R_α is a direct sum of isomorphic minimal right $\hat{R}$-ideals of $\hat{R}$, $\hat{R}_\alpha = \oplus\{\hat{U} \mid U \in I \cap \alpha\}$.

Even though the next corollary seems to be already known, it will nevertheless be stated because it follows immediately from the previous more general theory for modules. Although all of the previous results could be proved for a ring with or without an identity, the proof of (ii) and (iii) below actually requires that $1 \in R$.

4.8 COROLLARY 2 TO THEOREM III. Suppose that R is a ring with finite right uniform dimension $udR = m < \infty$ and with zero-singular right ideal $ZR = 0$. Then the following hold.

(i) $|\Omega| \equiv n \leq m$; $\Omega = \{\alpha(1),\ldots,\alpha(n)\}$ where n is the number of distinct isomorphy classes of injective hulls of uniform right ideals of R.

(ii) $|I \cap \alpha(j)| \equiv n(j) < \infty$ is finite, $1 \leq j \leq n$.

(iii) $\hat{R} = \hat{R}_1 \oplus\ldots\oplus \hat{R}_j \ldots \oplus\hat{R}_n$; $\hat{R}_j \lhd \hat{R}$, $j = 1,\ldots n$; $\hat{R}_j = \oplus\{\hat{U} \mid U \in I \cap \alpha(j)\}$ is a simple ring with the descending chain condition, where each $\hat{U} \subset \hat{R}$ is a minimal right $\hat{R}$-ideal of $\hat{R}$.

The next immediate objective is to formulate conditions under which indecomposable injectives U appearing previously in $\oplus\hat{U} << \hat{M}$ will be all isomorphic to one another.

The next definition of a prime submodule $K < M$ reduces to the usual one if $M = R$ and $1 \in R$ ([7; p.158]). To complicate matters, it is not the only definition with this property. Nevertheless, what does distinguish the next definition from the rest is that it is possible to develop the theory of primes in a more general module context, from which the usual facts when $K < M = R$ is a prime right (or two sided) ideal become mere corollaries, a special case within a theory having broader coverage ([7]).

4.9 DEFINITION. A submodule $K < M$ is <u>prime</u> if for any $\beta \in M\backslash K$ and $t \in R$, if $\beta Rt \subset K$, then $Mt \subset K$.

The module M is prime if $(0) < M$ is a prime submodule.

Given two submodules of the same module, the previous definition does not tell us which of the two submodules is more prime and which is less prime. The next flexible definition of primeness corrects this deficiency. The bigger the ring Λ below is, the "more prime" a submodule is.

4.10 DEFINITION. For any module M and right R-submodule $K < M$, suppose that Λ is some given subring $\Lambda \subset \mathrm{End}\hat{M}$ of R-homomorphisms $M \longrightarrow \hat{M}$. Then $K < M$ will be said to be <u>prime</u> with respect to Λ, if for any right Λ-ideal $A \subset \Lambda$ and for any right R-submodule $B < M$ if $AM \not\subset K$ but $AB \subset K$, then $\Longrightarrow B \subset K$.

As usual, M is a prime Λ-module if $(0) < M$ is a Λ-prime submodule.

4.11 CONSEQUENCES

(1) M is Λ-prime $\Longrightarrow \Lambda$ is a prime ring.

(2) For any right ideal $L < R$, take $K = L$, $M = R$, and $\Lambda = R$. Then $L < R$ is prime with respect to $\Lambda = R$ if and only if $L < R$ is prime in the ordinary usual sense (4.1), i.e. whenever $aRb \subset L$ with $a \notin L$, then $b \in L$.

The analogy between torsion free uniform modules and simple modules is visible in the equivalence of (i) and (ii) below.

4.12 THEOREM IV. For a module M satisfying 3.1, the following four conditions are all equivalent.

(i) All uniform submodules of $\hat{M}$ have isomorphic injective hulls.

(ii) Every fully invariant submodule of $\hat{M}$ is essential in $\hat{M}$.

(iii) M is $\mathrm{End}\hat{M}$-prime.

(iv) $\mathrm{End}\hat{M}$ is a prime ring.

4.13 REMARKS. The last theorem raises two questions or problems.

(1) Is there a fifth condition equivalent to (i)-(iv) for a module M prime in the ordinary sense of 4.9?

(2) In condition (ii), when can $\hat{M}$ be replaced by M?

For emphasis, it seems worthwhile to rephrase the objectives of the last theorem in the next corollary.

4.14 COROLLARY 1 TO THEOREM IV. If (i)-(iv) hold in the last theorem, then $\hat{M} \cong E(\oplus_I Y)$ for I-copies of the same indecomposable injective summand $Y < \hat{M}$ of $\hat{M}$.

The special case when $|I| < \infty$ of the next corollary is already in the literature ([8; p.328, 4.17] and [9; p.39, 2.14]). Note that conclusion (i) below also says that the endomorphism ring of each $\tilde{U}_i$ as well as $\hat{U}_i$ is a division ring.

4.15 COROLLARY 2 TO THEOREM IV. Suppose that $L < R$ is a prime complement right ideal with $ZR \subset L$ such that every nonzero submodule of R/L contains a uniform one. Then there exist

(i) prime, uniform, torsion free right ideals $U_i < R$ indexed by $i,j \in I$ such that

(ii) $\displaystyle\bigoplus_{i\in I} \frac{U_i \oplus L}{L} << \frac{R}{L}$; $L \oplus (\bigoplus_{i\in I} U_i) << R.$

(iii) $E(R/L) \cong E(\bigoplus_{i\in I} U_i) \subset \hat{R}$, where $\hat{U}_i \cong \hat{U}_j$ for all $i,j \in I$.

5. ENDOMORPHISM RINGS

The descriptions of M and $\hat{M}$ given by Theorems II and III are used in this section to completely determine the endomorphism ring of the injective hull $\hat{M}$ of the module M. Although these results in Section 5 are already known ([2], [16], and [17]), here it is shown how they are easy and natural consequences of the more systematic general theory in Section 3. Sometimes the results given here are more precise than in the literature (see 5.6).

The proofs of several of the implications in Theorem IV follow immediately from the complete description of $\mathrm{End}\hat{M}$ in 5.5 in this section. However, the shortest proof of Theorem IV does not require a complete explicit description of $\mathrm{End}\hat{M}$.

5.1 NOTATION. Throughout this section it will be assumed that the module M satisfies the hypothesis 3.1, i.e. contains an essential direct sum of torsion free uniform submodules $\oplus\{U \mid U \in I\} << M$. The main results of this section will be obtained from the characterization of $U\hat{M}$ in Proposition 3.13 as $U\hat{M} = \oplus\{(U\hat{M})_\alpha \mid \alpha \in \Omega\} = \oplus\{\hat{U} \mid U \in I\}$, where $(U\hat{M})_\alpha = (EM)_\alpha = \oplus\{\hat{U} \mid U \in I \cap \alpha\}$. Any right R-module -- such as $U\hat{M}$ or $(EM)_\alpha = (U\hat{M})_\alpha$ -- will be regarded as a right R and a left module over its R-endomorphism ring $\mathrm{End}\,U\hat{M}$ or $\mathrm{End}(EM)_\alpha$.

Since $U\hat{M}$ is quasi-injective (by 3.13 (iii)) with $UM \subset U\hat{M} \subset \hat{M}$, for any $\Phi \in \mathrm{End}\hat{M}$, $\Phi U\hat{M} \subset U\hat{M}$. Thus there is a ring homomorphism $\mathrm{End}\hat{M} \longrightarrow \mathrm{End}U\hat{M}$, $\Phi \longrightarrow \Phi|U\hat{M}$, where the latter is the restriction and corestriction. If $\Phi U\hat{M} = 0$, then since $U\hat{M} << \hat{M}$, also $\phi^{-1}0 << \hat{M}$, and $Z(\hat{M}/\Phi^{-1}0) = \hat{M}/\Phi^{-1}0 \cong \Phi\hat{M} \subset \hat{M}$ show that $\Phi = 0$. Thus $\mathrm{End}\hat{M} \cong \mathrm{End}U\hat{M}$. The injectivity of $\hat{M}$ trivially guarantees that the map is onto.

The use of 3.13 (iv) shows immediately that $\mathrm{End}U\hat{M} = \Pi\{\mathrm{End}(EM)_\alpha \mid \alpha \in \Omega\}$. Thus in order to describe or characterize $\mathrm{End}\hat{M} \cong \mathrm{End}U\hat{M}$, it will now suffice to determine the ring $\mathrm{End}(EM)_\alpha$.

5.2. A "full right linear ring" in the sense of [16; p.92] is a left one according to the usage in [2; p.250 and p.260 in 3.9]. Here this term will not be used, but rather a <u>full linear ring</u> of a <u>right vector space</u> will refer to the ring of all linear transformations written on the <u>left</u> of a <u>right</u> vector space over some division ring.

Let $Y \in I \cap \alpha$ be any fixed module and define $\Delta = \mathrm{End}\hat{Y}$. Since $ZY = 0$, $\hat{Y}$ is strongly uniform (or monoform), and hence Δ is a division ring. Define V to be a <u>right</u> vector space over Δ of right dimension $[V:\Delta] = |I \cap \alpha|$. Let $\mathrm{Hom}_\Delta(V,V)$ denote the ring of all Δ-linear transformations of V written on the left of V.

5.3. The right Δ-vector space V is isomorphic to all column vectors indexed by $I \cap \alpha$ with a finite number of nonzero components or entries from Δ. Then $\mathrm{Hom}_{\Delta}(V,V)$ is given by all so-called column finite square matrices $\|d_{ij}\|$ indexed by $i,j \in I \cap \alpha$ with entries in $d_{ij} \in \Delta$, where $\|d_{ij}\|$ acts by left multiplication on column vectors. Note that $|\{d_{ij} \mid i \in I \cap \alpha\}| < \infty$ for any fixed $j \in I \cap \alpha$, i.e. each column of $\|d_{ij}\|$ has at most a finite number of nonzero entries. However, $\|d_{ij}\|$ may not only have an infinite number of nonzero columns but even an infinite number of nonzero entries in some row.

In order to understand the present special case in a more inclusive framework, some general facts are listed for the reader's benefit. In particular, some of these facts would have to be used in order to prove the next theorem which describes the endomorphism ring of $\hat{M}$ for M as in 3.1.

5.4 FACTS. (1) For any injective right R-module W, set $S = \mathrm{End}W$. Then the Jacobson radical of S is $J(S) = \{\phi \in S \mid \ker \phi << W\}$, while $S/J(S)$ is a self right injective regular ring ([21; p.102, Proposition 1] and [16; p.52, Theorem 2.21]).

(2) If W is uniform and injective (with $ZW \neq 0$ or $ZW = 0$), then $S = \mathrm{End}W$ is a local ring. Hence $S/J(S)$ is a division ring ([21; p.104, Corollary 2]).

(3) If U is any uniform module with $ZU = 0$, then $\mathrm{End}\hat{U} \cong \mathrm{End}\tilde{U}$ is a division ring ([16; p.50, Corollary 2.17]).

(4) Let V be a right vector space over a division ring Δ. Then the full linear ring of a right vector space $S = \mathrm{End}V_{\Delta}$ is

(i) regular;

(ii) right self injective.

(iii) Furthermore, ${}_SS$ is left S-self injective if and only if the right Δ-dimension of V is finite, i.e. $[V:\Delta] \neq \infty$ (see [16; p.52-53, Proposition 2.23]).

(5) If R is commutative, and U is a torsion free uniform R-module, then $\hat{U} \cong \tilde{U}$.

A full proof of the next theorem would start out as in 5.1 with $\mathrm{End}\hat{M} \cong \mathrm{End}U\hat{M}$, and would be very long. The method of proof would be similar to [18; p.15-18], thus once again underscoring the analogy between torsion free uniform modules and simple modules.

5.5 THEOREM IV. Let M be any module containing an essential direct sum of torsion free uniform submodules with

I, $\alpha \in \Omega$, $U\hat{M} = \oplus\{(EM)_\alpha \mid \alpha \in \Omega\}$, Δ, V, $[V:\Delta]$, $\mathrm{Hom}_\Delta(V,V)$, and $\|d_{ij}\| \in \mathrm{Hom}_\Delta(V,V)$ as before (1.2, 3.1, 5.2, and 5.3). Then the following hold.

(i) $\mathrm{End}\hat{M} \cong \Pi\{\mathrm{End}(EM)_\alpha \mid \alpha \in \Omega\}$ where each

(ii) $\mathrm{End}(EM)_\alpha$ is a full linear ring of a right vector space.

(iii) $\mathrm{End}(EM)_\alpha \cong \mathrm{Hom}_\Delta(V,V)$, where $[V:\Delta] = |I \cap \alpha|$, and

$$\mathrm{Hom}_\Delta(V,V) \cong \left\{ \|d_{ij}\| \;\middle|\; \begin{array}{l} d_{ij} \in \Delta;\ i,j \in I \cap \alpha \\ \text{column finite} \end{array} \right\}$$

5.6 REMARKS. (1) The proof of the last theorem starts from $\mathrm{End}\hat{M} \cong \mathrm{End}U\hat{M}$, where once the notation is chosen (nonuniquely), $U\hat{M}$ can be made uniquely in an obvious way into a left vector space ${}_\Delta U\hat{M}$ over Δ in such a manner that R becomes a ring of Δ-linear transformations acting on the right of ${}_\Delta U\hat{M}$. For this reason it is natural to somehow jump to the erroneous conclusion that therefore $\mathrm{End}\hat{M}$ should be a full linear ring of a left vector space, i.e. $\mathrm{End}\hat{M} = \mathrm{Hom}_\Delta({}_\Delta UM, {}_\Delta UM)$. As can be clearly seen from the last theorem part (iii) and 5.4 (4) (iii), if $[V:\Delta] = \infty$, then there does NOT exist any division ring D and left D vector space ${}_D W$ such that $\mathrm{End}\hat{M} \cong \mathrm{Hom}_D({}_D W, {}_D W)$.

(2) Occasionally in the literature it is merely shown that $\mathrm{Hom}_R(EM,EM)$ or $\hat{R}$ is a direct product of endomorphism rings of vector spaces over division rings. In view of (1) this is an incomplete result which does not determine the ring in question.

5.7. As usual, a <u>minimal complement</u> submodule is a nonzero complement submodule containing no nonzero smaller complement submodule. The set of complement submodules of a torsion free module is partially ordered by inclusion with $\{0\}$ as its smallest element. This set is closed under arbitrary intersections. Thus an atom in this partially ordered set is a minimal complement submodule.

5.8. Let N be any right R-module with $ZN = 0$ and let $C < N$ be a complement submodule. Then the following hold.

(i) C is a minimal complement submodule $\Longleftrightarrow$ C is uniform.

(ii) Every nonzero submodule of N contains a uniform submodule (i.e. N satisfies 3.1) $\Longleftrightarrow$ every nonzero complement of N contains a miminal complement.

A known result ([16; p.93, 3.29] now follows in a very natural sort of way from the last theorem.

5.9 COROLLARY TO THEOREM IV. For a ring R with $ZR = 0$, the following three conditions are equivalent.

(i) Every nonzero right ideal of R contains a uniform right ideal.

(ii) The ring R is isomorphic to a direct product of full linear rings on right vector spaces.

(iii) Every nonzero complement submodule of M contains a minimal complement submodule.

PROOF. If $ZR = 0$, then $\hat{R} = ER$ is a ring. If M is a torsion free right R-module, then $\hat{M}$ is a right ER-module. Clearly, $\mathrm{Hom}_{ER}(\hat{M},\hat{M}) \subset \mathrm{Hom}_R(\hat{M},\hat{M})$. Conversely for any $\Phi \in \mathrm{Hom}_R(\hat{M},\hat{M})$, $\xi \in \hat{M}$, and $0 \neq x \in ER$, the element $y \equiv \Phi(\xi x) - (\Phi\xi)x \in \hat{M}$ regarded merely as an R-module satisfies $y(x^{-1}R) = 0$. Since $x^{-1}R << R$, $y \in Z\hat{M} = (0)$. Thus $\mathrm{Hom}_{ER}(\hat{M},\hat{M}) = \mathrm{Hom}_R(\hat{M},\hat{M}) \equiv \mathrm{End}\hat{M}$.

(i) $\Longrightarrow$ (ii). Since $1 \in ER$, $\mathrm{Hom}_{ER}(ER,ER) = ER$. Consequently in the special case when $M = R$, there is a ring isomorphism $\hat{R} \cong \mathrm{Hom}_R(\hat{R},\hat{R}) \equiv \mathrm{End}\hat{R}$, where by Theorem IV the latter is as required in (ii).

(ii) $\Longrightarrow$ (i). For any given right ideal $0 \neq A < R$, $\hat{A} \subset \hat{R}$ is a right $\hat{R}$-ideal of $\hat{R}$. A matrix computation shows that $\hat{A}$ contains a minimal $\hat{R}$-right ideal $0 \neq Y \subset \hat{A}$ of $\hat{R}$. The fact that Y is an $\hat{R}$-simple module guarantees that $0 \neq Y \cap R < \hat{R}$ is a uniform right R-ideal. If all modules are now regarded as R-modules, then since $0 \neq A << \hat{A}$ and since $0 \neq Y \cap R < \hat{A}$, it follows that $0 \neq Y \cap R \cap A < A$ is a nonzero uniform submodule of A.

The equivalence (iii) $\Longleftrightarrow$ (i) holds more generally (for modules) in 5.8.

One of the main motivations behind this note was to generalize the notion of the socle by replacing simples with uniform modules. The next remark shows that in the special case when $ZR = 0$, there is a very direct relation between uniform R and simple R-modules.

5.10 REMARK. If $V < N$ are any right modules over a ring R with both $ZN = 0$ and $ZR = 0$, then the following hold.

(1) For any $x \in N$, $E(x\hat{R}) = x\hat{R}$ (16; p.43, 2.6; p.45, 2.9]).

(2) V is uniform $\Longleftrightarrow$ $\hat{V}$ is a simple right $\hat{R}$-module (16; p.90, 3.24; p.43, 2.6; p.45, 2.9]).

(3) N satisfies 3.1 $\Longleftrightarrow$ $\hat{N}$ as a right $\hat{R}$-module has an essential socle ([16; p.91, 3.27], see also [16; p.18; p.42; p.43, 2.4]).

REFERENCES

[1] R. T. Bumby, Modules which are isomorphic to submodules of each other, Archiv. Math. XVI (1965), 1965), 184-185.

[2] S. Chase and C. Faith, Quotient rings and direct products of full linear rings, Math. Z. 88 (1965), 250-264.

[3] J. Dauns, Simple modules and centralizers, T.A.M.S. 166 (1972), 457-477.

[4] J. Dauns, One sided prime ideals, Pac. J. Math. 47 (1973), 401-412.

[5] J. Dauns, Quotient rings and one sided primes, J. Reine Angew. Math. 278/279 (1975), 205-224.

[6] J. Dauns, Generalized monoform and quasi-injective modules, Pac. J. Math. 66 (1976), 49-65.

[7] J. Dauns, Prime Modules, J. Reine Angew. Math. 298 (1978), 156-181.

[8] J. Dauns, Prime modules and one sided ideals, Algebra Proceedings III, University of Oklahoma, 1979, Marcel Dekker, p.41-83.

[9] J. Dauns, Uniform modules and complements, Houston J. Math. 6 (1980), 31-40.

[10] C. Faith and Y. Utumi, Quasi-injective modules and their endomorphism rings, Arch. Math. 15 (1964), 166-177.

[11] L. Fuchs, On primal ideals, P.A.M.S. 1 (1950), 1-6.

[12] L. Fuchs, On subdirect unions, I, Acta Scient. Math. III (1952), 103-120.

[13] L. Fuchs, On a new type of radical, Acta Scient. Math. XVI (1955), 43-53.

[14] L. Fuchs, Abelian p-Groups and Mixed Groups, Séminaire de Math. Sup., Les Presses de l'Université de Montreal.

[15] A. W. Goldie, Semi-prime rings with maximum condition, Proc. London Math. Soc. 10 (1960), 201-220.

[16] K. Goodearl, Ring Theory, Marcel Dekker, New York, 1976.

[17] R. Gordon, Krull Dimension, A.M.S., Memoirs No.133, Providence, R.I., 1973.

[18] J. Jans, Rings and Homology, Holt, Rinehart and Winston, New York, 1964.

[19] K. Koh, On some characteristic properties of self injective rings, P.A.M.S. 19 (1968), 209-213.

[20] K. Koh, Quasi-Simple Modules, Lectures on Rings and Modules, Lecture Notes in Mathematics No.246, Springer (1972), New York.

[21] J. Lambek, Lectures on Rings and Modules, Chelsea Publ. Co., New York 1976.

[22] L. Levy, Unique subdirect sums of prime rings, T.A.M.S. 106 (1963), 64-76.

[23] E. Matlis, Injective modules over Noetherian rings, Pac. J. Math. 8 (1958), 511-528.

[24] H. Storrer, On Goldman's Primary Decomposition, Lecture Notes in Mathematics No.246, Springer (1972), New York.

[25] M. Teply, Torsion free injective modules, Pac. J. Math. 28 (1969), 441-453.

[26] M. Teply, Subidealizer rings and the splitting properties, to appear.

[27] R. B. Warfield, Decomposition of injective modules, Pac. J. Math. 31 (1960), 236-276.

ON CENTRAL POLYNOMIALS AND ALGEBRAIC ALGEBRAS

Yehiel Ilamed
Soreq Nuclear Research Centre
Yavne 70600, Israel

1. INTRODUCTION

Let C be a commutative ring and let $C\{X\} = C\{x_1,x_2,\cdots\}$ denote the associative algebra over C generated by $X = \{x_1,x_2,\cdots\}$, a set of noncommuting variables. The elements of C{X} are polynomials in the variables x_1, with coefficients in C.

Let A be any algebra and let $p(a_1,\cdots,a_n)$ denote the image of an element $p(x_1,\cdots,x_n) \in C\{X\}$ under the homomorphism of C{X} into A sending $x_i \to a_i$, $i = 1,2\cdots$ We say that $p(x_1,\cdots,x_n)$ is an identity for A if $p(a_1,\cdots,a_n) = 0$ for all a_i in A. For example, Wagner ([1], p. 12) has shown that

$$(1) \qquad [x_1,x_2]^2x_3 - x_3[x_1,x_2]^2$$

is an identity for $M_2(C)$ the algebra of 2 by 2 matrices with entries in C. The symbol [,] denotes the commutator, $[x_1,x_2] = x_1x_2 - x_2x_1$.

A polynomial $p(x_1,\cdots,x_n) \in C\{X\}$ is called a central polynomial for an algebra A if $p(x_1,\cdots,x_n)$ is not an identity for A but

$$(2) \qquad [p(x_1,\cdots,x_n),\ x_{n+1}]$$

is an identity for A. For example, by identity (1) the polynomial

$$(3) \qquad [x_1,x_2]^2$$

is a central polynomial for $M_2(C)$.

We say that an algebra A is algebraic with respect to a subalgebra B of A if for each $a \in A$ there exists a polynomial f(x) with coefficients in B so that f(a) = 0; if B is noncommutative we distinguish between left and right polynomials f(x).

In this paper we prove (theorem 2) that if a homogeneous n-linear polynomial

$p(x_1,\cdots,x_n) \in C\{X\}$ is a central polynomial for an algebra A with identity 1 and if there exist $a_1,\cdots,a_n \in A$ so that $p(a_1,\cdots,a_n) = 1$, then the algebra A is algebraic with respect to Z(A) the center of A. For any $a \in A$ we give in theorem 2 the explicit expression of the coefficients (in Z(A)) of an n-th degree polynomial which has a as a root. The proof of theorem 2 is based on an identity (theorem 1) for C{X}. The case of an algebra without unity is discussed in Note 4. The results of this paper appeared in [2,3].

2. AN IDENTITY FOR ASSOCIATIVE ALGEBRAS

Let $p(x_1,\cdots,x_n)$ be defined by

(4) $$p(x_1,\cdots,x_n) = \Sigma_\sigma k_\sigma x_{\sigma 1} x_{\sigma 2} \cdots x_{\sigma n},$$

where the summation is over σ, the n! permutations of $1,2,\cdots,n$, and k_σ are n! given elements in C.

Let $f_j(p,x^{*j})$, $j = 0,1,\cdots,n$, be defined by

(5) $$f_j(p,x^{*j}) = \Sigma_{i1 + \cdots + in = j}\ p(x^{i1}x_1,\cdots,x^{in}x_n)$$

where the exponents $i1,\cdots,in$ are restricted to 0 and 1. For example

$f_0(p,x^{*0}) = p(x_1,\cdots,x_n)$, $f_n(p,x^{*n}) = p(xx_1,\cdots,xx_n)$ and

(6)

$$f_1(p,x^{*1}) = p(xx_1,x_2,\cdots,x_n) + p(x_1,xx_2,x_3,\cdots,x_n) + \cdots + p(x_1,\cdots,x_{n-1},xx_n).$$

Let $f_p(x_1,\cdots,x_n;x)$ be defined by

(7) $$f_p(x_1,\cdots,x_n;x) = \Sigma_{j=0}^{n}(-1)^j x^{n-j} f_j(p,x^{*j})$$

where $f_j(p,x^{*j})$, $j=0,1,\cdots,n$ are defined by Eqs. (4) and (5).

Theorem 1. The polynomial $f_p(x_1,\cdots,x_n;x)$ defined by Eq. (7) is an identity for C{X}.

Proof. We have to show that the right side of Eq. (7) is identically zero in the variables $x_1,\cdots,x_n$ and x. Let us define $f_{i,0}$ and $f_{i,1}$ by

$$(8)\qquad f_i(p,x^{*i}) = xf_{i,0} + f_{i,1},\ i = 0,1,\cdots,n\ ,$$

where $xf_{i,0}$ is the sum of all the monomials of $f_i(p,x^{*i})$ with x as their first factor and $f_{i,1}$ is the sum of all the other monomials.

It is clear from Eq. (6) that

$$(9)\qquad f_{0,0} = 0,\ f_{n,1} = 0 \text{ and } f_{0,1} = f_{1,0}.$$

Using Eqs. (5) and (8), we obtain

$$(10)\qquad f_{j,1} = f_{j+1,0},\ j = 0,1,\cdots,n-1.$$

Substituting $xf_{i,1} + f_{i,0}$ for $f_i(p,x^{*i})$ in Eq. (7), we obtain

$$(11)\qquad f_p(x_1,\cdots,x_n;x) = \Sigma_{j=0}^{n}(-1)^j x^{n-j}(xf_{j,0} + f_{j,1}).$$

Using Eqs. (9) and (10) we obtain $f_p(x_1,\cdots,x_n;x) = 0$.

3. THE MAIN THEOREM

Theorem 2. Let $p(x_1,\cdots,x_n) \in C\{X\}$, defined by Eq. (4), be a central polynomial for an algebra A with 1, assume that C is the ring of integers and assume that there exist $a_1,\cdots,a_n \in A$ so that

$$(12)\qquad p(a_1,\cdots,a_n) = 1.$$

Then each element $a \in A$ is a root of the polynomial

$$(13)\qquad h(x) = x^n - c_1x^{n-1} + c_2x^{n-2} + \cdots + (-1)^{n-1}c_{n-1}x + (-1)^n c_n$$

where the coefficients $c_i = c_i(a_1,\cdots,a_n;a)$ defined by (eq. 4))

(14) $$c_i(a_1,\cdots,a_n;a) = f_i(p(a_1,\cdots,a_n),a^{*i}),\ i = 1,\cdots,n,$$

are in the center of A.

Proof. Since $p(x_i,\cdots,x_n)$ is a central polynomial for A, it follows from Eqs. (5) and (14) that $c_i \in Z(A)$, $i = 1,\cdots,n$. Using theorem 1 and Eq. (12) we obtain $h(a) = 0$. Let us note that $c_i a = ac_i$.

4. NOTES

1. Let $p(x_1,\cdots,x_n)$ be defined by Eq. (4). If $p(x_1,\cdots,x_n)$ is a central polynomial for an algebra A and if $\Sigma_\sigma k_\sigma = k$ is an invertible element in C, then the n-th power of each element of A is in Z(A). Hence A is a K-ring [4], p. 218).

In the case $k = 0$ we cannot use this note, but we can use theorem 2.

2. Another proof of theorem 1 is as follows. It is trivial to prove theorem 1 if $p(x_1,\cdots,x_n) = k_1x_1x_2\cdots x_n$ or if $p(x_1,\cdots,x_n) = k_\sigma x_{\sigma 1}x_{\sigma 2}\cdots x_{\sigma n}$, a permutation of the factors of $x_1x_2\cdots x_n$. The identity (7) is the sum of the identities obtained for each monomial separately.

3. If $p(x_1,\cdots,x_n)$ is defined by Eq. (4) and if we define

(15) $$g_j(p,x^{*j}) = \Sigma_{i1+\cdots+in=j}\, p(x_1x^{i1},\cdots,x_nx^{in}),\ j = 0,1,\cdots,n\ ,$$

where the exponents $i1,\cdots,in$ are restricted to 0 and 1, then the polynomial $g_p(x_1,\cdots;x_n;x)$ defined by

(16) $$g_p(x_1,\cdots,x_n;x) = \Sigma_{j=o}^{n}(-1)^j g_j(p,x^{*j})x^{n-j}$$

is a polynomial identity for C{X}.

4. If $p(x_1,\cdots,x_n)$ is a central polynomial for an algebra A without 1, then each $a \in A$ is a root of the polynomial $h(x)$ defined by

(17) $$h(x) = x^n p(a_1,\cdots,a_n)-x^{n-1}c_1+x^{n-2}c_2+\cdots+(-1)^{n-1}xc_{n-1}+(-1)^n c_n$$

where $c_i = c_i(a_1,\cdots,a_n;a)$ are defined by Eq. (14) and $p(a_1,\cdots,a_n) \neq 0$.

The polynomial $h(x)$ is nontrivial since by assumption $p(x_1,\cdots,x_n)$ is central for A and hence there exist at least an n-list $a_1,\cdots,a_n$ in A so that $p(a_1,\cdots,a_n) \neq 0$.

5. If for an algebra A with 1, all the evaluations $p(a_1,\cdots,a_n)$ (p is defined by Eq. (4)) for $a_1,\cdots,a_n \in A$ are in a noncommutative subalgebra B of A and if there exist $a_1^0,\cdots a_n^0 \in A$ so that $p(a_1^0,\cdots,a_n^0) = 1$, then the identities (7) and (16) define, respectively, in a natural way (like in theorem 2) a right and a left polynomial with coefficients in B so that any $a \in A$ is a root of these polynomials.

It should be noted that here as well as in Eqs. (13) and (17) the coefficients of the polynomial are functions of the root a.

6. Let $p(x_1,\cdots,x_{2n})$ be defined by

$$(18)\qquad p(x_1,\cdots,x_{2n}) = \Sigma_\sigma \Gamma_\tau (sg\sigma)(sg\tau) x_{\sigma 1} x_{n+\tau 1} x_{\sigma 2} x_{n+\tau 2} \cdots x_{\sigma n} x_{n+\tau n}$$

where Σ_σ means summation over all the n! permutations σ of $1,\cdots,n$, Γ_τ means summation over all the n cyclic permutations τ of $1,\cdots,n$ and sg means sign. The polynomial (18) is a central polynomial for $M_k(C)$ if $n = k^2$, [5].

If $p(x_1,\cdots,x_{2n})$ defined by Eq. (18) for $n = k^2$ is a central polynomial for an algebra A with 1 and if there exist $a_1,\cdots a_{2n}$ in A so that $p(a_1,\cdots,a_{2n}) = 1$, then each $a \in A$ is a root of

$$(19)\qquad h(x) = x^{2n} + \Sigma_{j=1}^{2n} (-1)^j c_j(a_1,\cdots,a_{2n};a) x^{n-j}$$

where the coefficients c_j are defined by Eq. (5) and (18).

It should be noted that the algebra A is algebraic, with respect to its center, of degree bounded by $2k^2$ while the algebra $M_k(C)$ is algebraic, with respect to C, of degree bounded by k.

REFERENCES

[1] N. Jacobson, PI-Algebra, An Introduction, Lecture Notes in Math., Vol. 441, Springer-Verlag, Berlin (1975).

[2] Y. Ilamed, On central polynomials and algebraic algebras, I and II, Notices Amer. Math. Soc., 26 (1979), A201 and A266.

[3] Y. Ilamed, On identities for associative algebras, Notices Amer. Math. Soc., 26 (1979), A421.

[4] N. Jacobson, Structure or rings, Colloquium Publications, Vol. XXXVII, Amer. Math. Soc., Providence (1968).

[5] S.A. Amitsur, On a central identity for matrix rings, J. London Math. Soc., (2), 14 (1976), 1-6.

FLATNESS AND f-PROJECTIVITY OF TORSION-FREE MODULES AND INJECTIVE MODULES

Marsha Finkel Jones

University of North Florida
Jacksonville, FL 32216

Let R be a ring with identity. Chase [5] has characterized the right coherent rings; he has shown that arbitrary products of $_RR$ are flat if and only if finitely generated submodules of free right R-modules are finitely presented. As compared to considering flatness of products of a projective generator, this paper investigates flatness (and related properties) of products of an injective cogenerator.

This condition turns out to be intimately related to several other properties of independent interest. One is the condition that torsion-free or injective modules are flat or f-projective (the latter is a property stronger than flatness); see, e.g., [6], [8] and [26]. Another is the condition that finitely presented (FP) or finitely generated (FG) modules embed in free modules; see, e.g., [12].

We approach these topics via torsion theory, but this paper requires a minimum of torsion theory background. (See Section 1.) Let $(\mathcal{T},\mathcal{F})$ be an hereditary torsion theory for right R-modules, and let E_R be an injective that cogenerates $(\mathcal{T},\mathcal{F})$. We first show (Proposition 2.1) that if R is torsion free then arbitrary products of E_R are flat if and only if every FP (relative to $(\mathcal{T},\mathcal{F})$), torsion-free M_R

embeds in a free module if and only if every torsion-free injective is flat. This easily yields characterizations of rings for which every torsion-free M_R is flat (Theorem 2.3) and of rings for which every injective is flat (Theorem 2.2).

We say that M is f-projective if, for each FG $N \leq M$, the inclusion map factors through a FG free module. (See [7], [14], [15] or [22].) (Then projective $\overset{\Rightarrow}{\nLeftarrow}$ f-projective $\overset{\Rightarrow}{\nLeftarrow}$ flat [15].) We show (Proposition 2.9) that arbitrary products of E_R are f-projective if and only if every FG torsion-free M_R embeds in a free module if and only if every torsion-free injective M_R is f-projective. This easily yields characterizations of rings for which every injective is f-projective (Theorem 2.10) and of rings for which every FG torsion-free M_R is projective (Theorem 2.11).

In Theorem 3.1 we show that a right nonsingular ring R with right maximal quotient ring $Q = Q_r$ has the unexpected property that Q_R is flat (f-projective) if and only if arbitrary products of Q_R are flat (f-projective). When combined with the earlier results, this easily yields a number of applications to nonsingular rings. For example, every FG nonsingular M_R is projective precisely when Q_R is f-projective and every FG torsionless M_R $({}_RX)$ is projective (Theorem 3.6).

§1. Preliminaries

Throughout this paper, all rings are associative with identity and all modules M are (unitary) right R-modules unless otherwise stated. The term mapping indicates an R-homomorphism; the symbol M_R denotes the category of right R-modules. The reader should consult [1] for background in ring theory.

A subclass $\mathcal{T}$ of M_R is called an <u>hereditary torsion class</u> if it is closed under submodules, homomorphic images, extensions and direct sums. The class $\mathcal{T}$ uniquely determines a <u>torsion-free class</u> $\mathcal{F}$ that is closed under injective hull, submodules, extensions and direct products; the pair $(\mathcal{T},\mathcal{F})$ is called an <u>hereditary torsion theory</u> for right R-modules. We say that an injective module E <u>cogenerates</u> the hereditary torsion theory $(\mathcal{T},\mathcal{F})$ if $\mathcal{F} = \{M_R | M \text{ embeds in } E^I \text{ for some set } I\}$.

Let t(M) denote the largest submodule of M that is in $\mathcal{T}$. Then M is <u>torsion</u> ($M \in \mathcal{T}$) if and only if t(M) = M, while M is <u>torsion free</u> ($M \in \mathcal{F}$) if and only if t(M) = 0. The <u>localization of M</u> with respect to $(\mathcal{T},\mathcal{F})$ (also called the module of quotients of M) is denoted by $L_{\mathcal{T}}(M)$. If $Q = L_{\mathcal{T}}(R)$ then there is a canonical ring homomorphism from R into Q; this mapping is a monomorphism precisely when $R \in \mathcal{F}$. For a more detailed torsion-theoretic background, see, e.g., Stenström [23] or Golan [10].

Let $(\mathcal{T},\mathcal{F})$ be any hereditary torsion theory for M_R, with corresponding torsion radical t. Then M_R is said to be <u>t-finitely generated</u> (t-FG) if $M/M' \in \mathcal{T}$ for some finitely generated (FG) $M' \leq M$; M_R is said to be <u>t-finitely presented</u> (t-FP) if there exists an exact sequence $0 \to K \to F \to M \to 0$) with F FG free and K t-FG.

<u>Remarks</u>. (1) If M is FG (respectively, FP) it is clearly t-FG (respectively, t-FP). Thus, for example, every FG free F is t-FP.

(2) If $\mathcal{T} = \{0\}$ then M is t-FG (t-FP) if and only if M is FG (FP).

(3) Every right torsion module is t-FG.

Both t-FG and t-FP modules are studied at length in [16], where they are used to characterize t-coherent rings [Theorems 3.3 and 3.4].

(We say that R is right (t-) coherent if every FG right ideal of R is (t-) FP. Note that (for $T = \{0\}$) 0-coherence coincides with the usual definition of coherence.)

In what follows we shall make frequent use of Lazard's characterization of flatness: a module M is flat if and only if every mapping from a finitely presented (FP) module into M factors through a FG free module [17].

§2. Flatness and f-Projectivity of Torsion-free or Injective Modules

If R is a right nonsingular ring, then the maximal right quotient ring $Q = Q_r$ of R is the localization of R with respect to the singular torsion theory. When viewed as a right R-module, however, Q_R is an injective hull of R and cogenerates the singular torsion theory. (See, e.g., Stenström [23].) We have previously shown [15, Proposition 2.1] that Q_R is f-projective if and only if Q_R is flat and $R \to Q$ is a ring epimorphism. Goodearl has shown [12, Theorem 2.4] that this is equivalent to the condition that every finitely generated nonsingular right R-module can be embedded in a free R-module. We shall show (Theorem 3.1) that arbitrary products of Q_R are f-projective if Q_R is f-projective. Thus every finitely generated nonsingular right R-module embeds in a free module if and only if products of an injective that cogenerates the torsion theory are f-projective.

This leads us to consider a more general situation. Let (T,F) be an arbitrary hereditary torsion theory, and let E_R be an injective that cogenerates (T,F). We shall consider the condition that arbitrary products of E_R are flat, i.e., the condition that E_R is Π-flat in the notation of Colby and Rutter [9]. We shall also consider the conditions that E_R is Π-R-ML or Π-f-projective (definitions below). As in the case of nonsingular rings, we will use finitely generated torsion-free modules in the characterization of the condition that E_R is Π-f-projective; the t-finitely presented torsion-free modules are of interest in the study of when E_R is Π-flat.

<u>Proposition 2.1</u>. Let (T,F) be an hereditary torsion theory for M_R and let E_R be an injective that cogenerates (T,F). Then (1) $\Rightarrow$ (2) $\Leftrightarrow$ (3) $\Leftrightarrow$ (4). If $R \in F$ then (4) $\Rightarrow$ (1).

(1) Every t-FP torsion-free M_R embeds in a free module.

(2) E_R is Π-flat.

(3) Every torsion-free injective M_R is flat.

(4) The injective hull of every (t-FP) torsion-free M_R is flat.

<u>Proof</u>. (1) $\Rightarrow$ (2). Suppose that (1) holds. Let I be any set and $f\colon P \longrightarrow E^I$, where P is FP. We want to show that f factors through a FG free module. Since $E^I \in F$, $t(P) \subseteq \mathrm{Ker}(f)$. Thus f induces $\bar{f}\colon P/t(P) \longrightarrow E^I$ such that $\bar{f}\eta = f$, where $\eta\colon P \longrightarrow P/t(P)$ is the natural map. Then triangle I of the following diagram commutes.

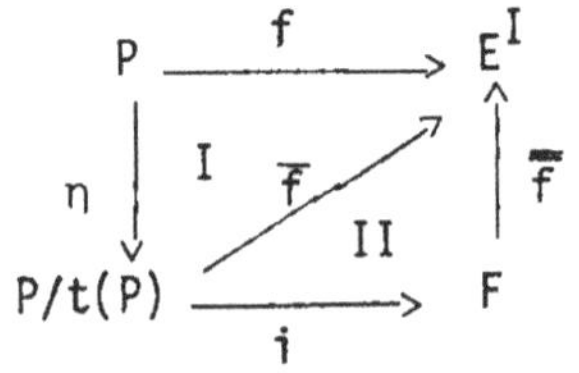

By [16, Lemma 2.4], P/t(P) is t-FP. Hence, by assumption, there is a free module F_R and a monomorphism $i\colon P/t(P) \longrightarrow F$. Since E^I is injective there is an $\bar{\bar{f}}\colon F \longrightarrow E^I$ such that II commutes. The conclusion follows from the fact that Im(i) is FG and hence is contained in a FG free $F' \subseteq F$.

(2) ⇔ (3). Suppose that (2) holds and let M_R be a torsion-free injective module. Then M embeds in E^I for some set I. The injectivity of M makes it a direct summand of E^I; hence M is flat. The converse is clear.

(3) ⇔ (4). Since $(\mathcal{T},\mathcal{F})$ is hereditary, M is torsion free if and only if the injective hull of M, I(M), is torsion free.

(4) ⇒ (1). Now assume that $R \in \mathcal{F}$, and suppose that (4) holds. Let M_R be torsion free and t-FP. By [16, Lemma 2.4] $M \cong P/t(P)$, where P is FP. Since M is torsion free, I(M) is a flat right R-module. Thus there is a FG free F_R and maps f, g such that the following diagram is commutative, where η and i are the canonical maps.

$$\begin{array}{ccccc} P & \xrightarrow{\ \eta\ } & P/t(P) \cong M & \xrightarrow{\ i\ } & I(M) \\ & {}_{f}\searrow & & \nearrow_{g} & \\ & & F & & \end{array}$$

Since $F \in \mathcal{F}$ by assumption, $t(P) \subseteq \operatorname{Ker}(f)$; f therefore induces $\bar{f}: M \longrightarrow F$ with $\bar{f}\eta = f$. Then $i = g\bar{f}$ and the conclusion follows from the fact that i is a monomorphism. □

<u>Remark</u>. In [16, Corollary 3.5] it is shown that if C_R is flat, then C_R is Π-flat if and only if R is left t_0-coherent, where $\mathcal{T}_0 = \operatorname{Ker}(C \otimes \_\_)$. Thus it is clear that (2) can be replaced by:

(2') E_R is flat and R is left t_0-coherent, where $\mathcal{T}_0 = \operatorname{Ker}(E \otimes \_\_)$.

<u>Theorem 2.2</u>. The following statements are equivalent for R.

(1) Every FP M_R embeds in a free module.

(2) Every injective M_R is flat.

(3) The injective hull of every FP M_R is flat.

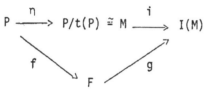

(4) If E_R is an injective cogenerator for M_R, then E_R is Π-flat.

(5) $R_R^+ = \mathrm{Hom}_{\mathbb{Z}}(R, \mathbb{Q}/\mathbb{Z})$ is flat and R is left t_0-coherent, where $T_0 = \mathrm{Ker}(R^+ \otimes \_\_)$.

Proof. Let $T = \{0\}$. Then the equivalence of (1) through (4) follows from the proposition. The equivalence of (4) and (5) then follows from [16, Corollary 3.5] and the observation that R_R^+ is an injective cogenerator for M_R. □

Remark. This theorem has also been proved by Colby [8, Theorem 1], who calls a ring satisfying these conditions a right IF ring. The development above is much simpler than in [8].

If we take the conditions of Proposition 2.1 and assume in addition that R has weak global dimension ≤ 1 ($\mathrm{WGD}(R) \leq 1$), we obtain the following theorem. (Cf., Chase [5, Theorem 4.1], Turnidge [24, Theorem 2.1], and Cheatham-Enochs [6, Theorems 1 and 2].)

Theorem 2.3. Let (T,F) be an hereditary torsion theory for M_R and let E_R be an injective that cogenerates (T,F). Then (1) $\Rightarrow$ (2) $\Rightarrow$ (3) $\Rightarrow$ (4). If $R \in F$, then (4) $\Rightarrow$ (1).

(1) E_R is Π-flat and R is left semihereditary.

(2) E_R is Π-flat and $\mathrm{WGD}(R) \leq 1$.

(3) The injective hull of every torsion-free M_R is flat and $\mathrm{WGD}(R) \leq 1$.

(4) Every torsion-free M_R is flat.

In particular, if (4) holds then $Q = L_T(R_R)$ is flat as a right R-module.

Proof. (1) $\Rightarrow$ (2). R is left semihereditary if and only if R is left coherent and $\mathrm{WGD}(R) \leq 1$ [5, Theorem 4.1].

(2) ⇒ (3). If (2) holds and $M_R \in F$, then $I(M) \in F$ and hence embeds in E^J for some set J. Since $WGD(R) \leq 1$, I(M) is flat.

(3) ⇒ (4). Clear.

(4) ⇒ (1). Now suppose that $R \in F$ and suppose that (4) holds. Then E_R^I is flat for every I. Now if M_R is FG torsionless, it is torsion free since $R \in F$; hence it is flat. Thus R is left semihereditary by Chase [5, Theorem 4.1]. □

Remark. If $M_R \in F$, then $M_R = \varinjlim P_i = \varinjlim P_i/t(P_i)$, where each P_i is FP. Thus any $M_R \in F$ is a direct limit of t-FP [16, Lemma 2.4] torsion-free modules. Now if $R_R \in F$, every t-FP flat M_R is projective [16, Lemma 3.9]. Thus, if $R \in F$, statement (4) may be replaced by:

(4') every t-FP torsion-free M_R is projective.

If we take the conditions of Theorem 2.3 and let $T = \{0\}$, the conditions collapse to the statement that every right R-module is flat; R is a regular ring. (Colby has shown [8, Proposition 5] that finite weak global dimension is enough to imply that R is regular.)

We shall now investigate the conditions that every FG (torsion-free) right R-module embeds in a FP module. To do this we must first discuss R-ML modules.

If M is an R-module and I is any set, then M^I denotes the product of card(I) copies of M, viewed as an R-module. Let $\mu_{M,I}: M \otimes R^I \longrightarrow M^I$ denote the canonical map $\mu_{M,I}(m \otimes \{r_i\}) = \{mr_i\}$. Lenzing [18] has shown that $\mu_{M,I}$ is an epimorphism (respectively, isomorphism) for every set I if and only if M is FG (respectively, FP). If, on the other hand, $\mu_{M,I}$ is a monomorphism for every I, Clarke [1] has called M an

R-Mittag Leffler (R-ML) module. Thus M is FP if and only if M is FG and R-ML.

Like flatness and f projectivity, the property of being R-ML can be characterized in terms of factorization of mappings. This has been done independently by both Goodearl [11, Theorem 1] and Clarke [7, Theorem 2.4].

Proposition 2.4. A module M is R-ML if and only if, for every FG submodule C of M, the inclusion map factors through a FP module.

With this characterization it is easy to show that a direct summand of an R-ML module is R-ML.

Proposition 2.5. Let $(\mathcal{T},\mathcal{F})$ be an hereditary torsion theory for right R-modules, and let E_R be an injective that cogenerates $(\mathcal{T},\mathcal{F})$. Then the following statements are equivalent.

(1) Every FG torsion-free M_R embeds in a FP module.

(2) E_R is Π-R-ML.

(3) Every torsion-free injective M_R is R-ML.

(4) The injective hull of every (FG) torsion-free M_R is R-ML.

Proof. (1) $\Rightarrow$ (2). Suppose that (1) holds. Let I be any set and let M_R be a FG submodule of E^I. We want to show that the inclusion map factors through a FP module. Since M is torsion free, $M \subseteq P$ for some FP P_R. The conclusion follows from the injectivity of E^I.

(2) $\Rightarrow$ (3). The proof of this implication is the same as for Proposition 2.1, noting that a direct summand of an R-ML module is R-ML.

(3) $\Rightarrow$ (4). Since $(\mathcal{T},\mathcal{F})$ is hereditary, $I(M) \in \mathcal{F}$ whenever $M \in \mathcal{F}$.

(4) $\Rightarrow$ (1). Suppose that (4) holds. Let M_R be FG and torsion free,

and let I(M) be the injective hull of M. Since I(M) is R-ML, the inclusion map i: $M \to I(M)$ factors through a FP P_R. Since i is a monomorphism, the conclusion follows. □

Theorem 2.6. The following statements are equivalent for R.

(1) Every FG M_R embeds in a FP module.

(2) Every injective M_R is R-ML.

(3) The injective hull of every FG M_R is R-ML.

(4) If E_R is an injective cogenerator for $\mathcal{M}_R$, then E_R is Π-R-ML.

Proof. Let $\mathcal{T} = \{0\}$ and apply the proposition. □

A module M is said to be f-projective [15] if, for every finitely generated (FG) submodule C of M, the inclusion map factors through a FG free module. (Simson [22] uses the term $\aleph_{-1}$-projective.) Clearly every projective module is f-projective, and every FG f-projective module is projective. In general, projective $\overset{\Rightarrow}{\nLeftarrow}$ f-projective $\overset{\Rightarrow}{\nLeftarrow}$ flat [15].

It is easy to show that a directed union of f-projective modules is f-projective and that a direct summand of an f-projective module is f-projective. Also, the combination of the Lazard and Goodearl characterizations of flat and, respectively, R-ML modules yields the following useful characterization of f-projectivity.

Proposition 2.7. (Clarke [7].) A module M is f-projective if and only if it is flat and R-ML.

An R-module M is called torsionless if it embeds in a direct product of copies of R. A ring R is said to be strongly right (t-) coherent if every FG torsionless right R-module is (t-) finitely presented [14].

In [14, Proposition 3.4] it is shown that if C_R is f-projective, then C_R is Π-f-projective if and only if R is strongly left t_0-coherent, where $\mathcal{T}_0 = \text{Ker}(C \otimes \_\_)$. (For $\mathcal{T}_0 = \{0\}$, take $C_R = R_R$.) Thus R is strongly right coherent if and only if ${}_R R$ is Π-flat and ${}_R R$ (or R_R [14, Lemma 3.13]) is Π-R-ML.

If we now take the conditions of Proposition 2.5 and assume in addition that R is right coherent, we obtain the following theorem. (Cf., [14, Proposition 3.4] and Goodearl [12, Theorem 2.4].)

Theorem 2.8. Let $(\mathcal{T}, \mathcal{F})$ be an hereditary torsion theory for $\mathcal{M}_R$ and let E_R be an injective that cogenerates $(\mathcal{T}, \mathcal{F})$. Then $1 \Rightarrow 2 \Rightarrow 4$ and $3 \Leftrightarrow 4$. If $R \in \mathcal{F}$ then $(4) \Rightarrow (1)$.

(1) E_R is Π-R-ML and R is strongly right coherent.

(2) E_R is Π-R-ML and R is right coherent.

(3) Every torsion-free M_R is R-ML.

(4) Every FG torsion-free M_R is FP.

In particular, if (4) holds then $Q = L_{\mathcal{T}}(R)$ is R-ML as a right R-module.

Proof. $(1) \Rightarrow (2)$. Clear, since f-projective implies flat.

$(2) \Rightarrow (4)$. Suppose that (2) holds and let M_R be FG torsion free. Then M_R embeds in a FP module by Proposition 2.5 and hence is FP since R is right coherent (e.g., [16, Theorem 3.3]).

$(3) \Leftrightarrow (4)$. Clear, since every module is a directed union of its FG submodules and FG + R-ML $\equiv$ FP.

$(4) \Rightarrow (1)$. Now assume that $R \in \mathcal{F}$ and (4) holds. Then E_R^I is R-ML for every I by Proposition 2.5. Since $R \in \mathcal{F}$, every FG torsionless M_R is torsion free and hence FP by (4). Thus R is strongly right coherent. □

If we take the conditions of Theorem 2.8 and let $T = \{0\}$, the conditions again collapse, this time to the statement that every FG right R-module is FP; R is right Noetherian.

Proposition 2.9. Let (T,F) be an hereditary torsion theory for right R-modules, and let E_R be an injective that cogenerates (T,F). Then the following statements are equivalent.

(1) Every FG torsion-free M_R embeds in a free module.

(2) E_R is Π-f-projective.

(3) Every torsion-free injective M_R is f-projective.

(4) The injective hull of every (FG) torsion-free M_R is f-projective.

In particular, if these conditions hold then $\{\text{FG torsion-free } M_R\} \subseteq \{\text{FG torsionless } M_R\}$. Thus if the conditions hold and $R \in F$, these two classes coincide.

Proof. Recall that a module X_R is f-projective if and only if, for every FG $Y_R \leq X_R$, the inclusion map factors through a FG free module. Thus the proofs of (1) $\Rightarrow$ (2), (3) $\Rightarrow$ (4) and (4) $\Rightarrow$ (1) parallel the proofs in Proposition 2.5. Since a direct summand of an f-projective module is f-projective, the proof of (2) $\Rightarrow$ (3) is the same as for Proposition 2.1. □

Remark. Since C_R is f-projective if and only if it is flat and R-ML (Proposition 2.7), we could also establish the implications (1) $\Rightarrow$ (2) $\Rightarrow$ (3) $\Rightarrow$ (4) by combining the conditions of Propositions 2.1 and 2.5.

Theorem 2.10. The following statements are equivalent for R.

(1) Every FG M_R embeds in a free module.

(2) Every injective M_R is f-projective.

(3) The injective hull of every FG M_R is f-projective.

(4) If E_R is an injective cogenerator for M_R, then E_R is Π-f-projective.

(5) $R_R^+ = \mathrm{Hom}_{\mathbb{Z}}(R,\mathbb{Q}/\mathbb{Z})$ is f-projective and R is left strongly t_0-coherent, where $T_0 = \mathrm{Ker}(R^+ \otimes \_\_)$.

Proof. The equivalence of (1) through (4) follows from Proposition 2.9, with $T = \{0\}$.

(4) $\Leftrightarrow$ (5). Follows from [14, Proposition 3.4] and the observation that ${}_R R^+$ is an injective cogenerator for M_R. □

Remark. The equivalence of (1), (2) and (4) could also be established by combining Theorems 2.2 and 2.6.

We include the following left-right symmetric theorem for the sake of completeness.

Theorem 2.11. [14, Theorem 3.15] The following statements are equivalent for a ring R.

(1) R is strongly right coherent and left semihereditary.

(2) Every torsionless right R-module is f-projective.

(3) Every finitely generated torsionless right R-module is projective.

(1') R is strongly left coherent and right semihereditary.

Remark. It is not difficult to show that R is left semihereditary precisely when submodules of f-projective left R-modules are f-projective. Hence the above theorem parallels the Chase result [5, Theorem 4.1] that R is right coherent and submodules of flat (left) R-modules are flat (i.e., WGD (R) $\leq$ 1) if and only if every (finitely generated) torsionless left R-module is flat (if and only if R is right semihereditary).

If we take the conditions of Proposition 2.9 and assume in addition that R is right semihereditary, we obtain the following theorem. (Cf., Theorem 2.11, Goodearl [12, Theorem 2.5] and Cateforis [3, Theorem 2.3].)

Theorem 2.12. Let (T,F) be an hereditary torsion theory for M_R, E_R an injective that cogenerates (T,F), and $Q = L_T(R_R)$. Then (1) $\Rightarrow$ (2) $\Rightarrow$ (3) $\Rightarrow$ (4) $\Leftrightarrow$ (5). If $R \in F$ then (5) $\Rightarrow$ (1).

(1) Every FG torsion-free M_R is torsionless and every FG torsionless M_R ($_RX$) is projective.

(2) E_R is Π-f-projective and R is strongly left and right coherent with WGD(R) ≤ 1.

(3) E_R is Π-f-projective and R is right semihereditary.

(4) Every torsion-free M_R is f-projective.

(5) Every FG torsion-free M_R is projective.

If (5) holds then Q_R is (Π-) f-projective. Hence Q_R is (Π-) flat, $R \to Q$ is a ring epimorphism, and $Q \cdot \ell_R(t(R)) = Q$.

Proof. (1) $\Rightarrow$ (2). If (1) holds then R is strongly left and right coherent and WGD(R) ≤ 1 by Theorem 2.11 and [5, Theorem 4.1]. In addition, every FG $M_R \in F$ is projective, so E_R^I, which is a directed union of its FG submodules, is f-projective for every set I.

(2) $\Rightarrow$ (3). Chase [5, Theorem 4.1].

(3) $\Rightarrow$ (4). If (3) holds and M_R is torsion free, then M_R is a submodule of an f-projective. Since R is right semihereditary, M_R is then f-projective.

(4) $\Rightarrow$ (5). A finitely generated f-projective module is projective.

(5) ⇒ (4). Clear, since a directed union of projective modules is f-projective.

(5) ⇒ (1). Now suppose that $R \in F$ and (5) holds. By (5), every FG torsion-free module is torsionless. Since $R \in F$, every FG torsionless M_R is torsion free and hence projective by (5). By Theorem 2.11 we also have FG torsionless $_RX$ projective.

Now if (5) holds it is clear that $Q = L_T(R_R)$ is right (Π-) f-projective and hence (Π-) flat. As Q_R is f-projective, $R \to Q$ is a ring epimorphism and $Q \cdot \ell_R(t(R)) = Q$ by [15, Theorem 2.4], where $\ell_R(t(R))$ denotes the left annihilator in R of t(R). □

Once again, if we take the conditions of Theorem 2.12 and let $T = \{0\}$, the conditions collapse to the statement that every finitely generated right R-module is projective; R is semisimple Artinian.

§ 3. Applications to Nonsingular Rings

We shall now turn our attention to applications of the preceding results to the case of the right maximal quotient ring Q_r of a right nonsingular ring. There is an extensive body of work in the literature on nonsingular rings. (See, for example, the bibliography in Goodearl [13]. See also [3], [4], [12], [13], [20], [21], [24] and [26].) Nevertheless, the coherence-type results of the preceding section, while aiding in the development of a more general framework for some of that work, also provide new information on nonsingular rings.

For the remainder of this paper we let R *be a right nonsingular ring with right maximal quotient ring* $Q = Q_r$. Then Q_R is an injective that cogenerates the singular torsion theory; we shall denote the corresponding torsion radical by Z. Recall that in this case every finitely generated nonsingular right R-module can be embedded in a finite direct sum of copies of Q [3, Lemma 2.2].

It is known ([15, Proposition 3.7] and [14, Theorem 3.7]) that $_RQ$ is flat (f-projective) if and only if $_RQ$ is Π-flat (Π-f-projective). It is not clear whether $_RQ$ R-ML implies $_RQ$ Π-R-ML.

Theorem 3.1. The following statements hold.

(1) If Q_R is flat then Q_R is Π-flat.

(2) If Q_R is R-ML then Q_R is Π-R-ML.

(3) If Q_R is f-projective then Q_R is Π-f-projective.

Proof. (1). Suppose that Q_R is flat and $f: P \to Q^I$, where P_R is FP. We want to show that f factors through a FG free module [17]. Since Im(f) is a FG submodule of Q^I, it is FG nonsingular and hence embeds in $\overset{n}{\underset{1}{\oplus}} Q$ for some n by [3, Lemma 2.2]. Since Q^I is injective as a right R-module, there is a map g such that triangle I of the following diagram commutes.

$$
\begin{array}{ccccc}
P & \xrightarrow{\ f\ } & \mathrm{Im}\,f & \overset{\subseteq}{\longrightarrow} & Q^I \\
f_1 \downarrow & \mathrm{II} & \subseteq \Big\downarrow \ \ {}^{I} & \nearrow & \\
F_R & \xrightarrow{f_2} & \underset{1}{\overset{n}{\oplus}} Q & {}^{g} & \\
\end{array}
$$

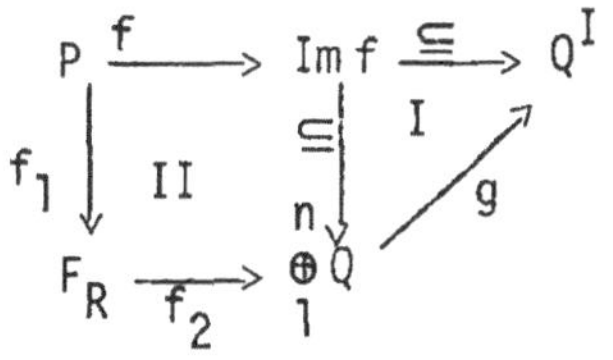

The flatness of $\overset{n}{\underset{1}{\oplus}} Q_R$ implies that there is a FG free F_R and mappings f_1 and f_2 such that II commutes. The conclusion follows from the diagram.

(2), (3). M_R is R-ML (f-projective) if and only if, for each FG submodule N of M, the inclusion map factors through a FP (FG free) right R-module. The proof proceeds as in (1). □

By combining Theorem 3.1 with the torsion-theoretic results of Section 2 we can show that: Q_R is flat precisely when every Z-FP nonsingular module embeds in a free module; Q_R is flat and WGD(R) ≤ 1 if and only if every (FG) nonsingular module is flat (see also [24, Theorem 2.1]); Q_R is R-ML precisely when every FG nonsingular module embeds in a FP module. (Use Proposition 2.1, Theorem 2.3 and Proposition 2.5, respectively.) Theorems 2.8 and 3.1 yield the following proposition.

Proposition 3.2. The following statements are equivalent.

(1) Every FG nonsingular M_R is FP.

(2) Q_R is (Π-) R-ML and R is strongly right coherent.

(3) Q_R is (Π-) R-ML and R is right coherent.

Corollary 3.3. If Q_R is R-ML and R is right coherent, then ${}_RQ$ is flat and $R \to Q$ is a ring epimorphism.

Proof. If the conditions hold, then every FG nonsingular M_R is FP, hence Z-FP. But ${}_RQ$ is flat and $R \to Q$ a ring epimorphism if and only if every FG nonsingular M_R is Z-FP [3, Theorem 1.6]. □

The equivalence of statements (1), (4) and (6) in the following proposition has also been shown by Goodearl [12, Theorem 2.4].

Theorem 3.4. The following statements are equivalent for the ring R.

(1) Every FG nonsingular M_R embeds in a free module.

(2) Q_R is (Π-) f-projective.

(3) Every nonsingular injective M_R is f-projective.

(4) Q_R is (Π-) flat and $R \to Q$ is a ring epimorphism.

(5) $Q \cdot Rq^{-1} = Q$ for every $q \in Q$, where $Rq^{-1} = \{r \in R: \ rq \in R\}$.

(6) $Q \cdot \bigcap_1^n Rq^{-1} = Q$ for every $q_1, \ldots, q_n \in Q$.

Proof. The equivalence of (1), (2) and (3) follows from Proposition 2.9 and Theorem 3.1, while that of (2), (4), (5) and (6) follows from Theorem 3.1, [15, Proposition 2.1], and the Popescu-Spircu theorem [19, Théorème 2.7]. □

Cateforis and Sandomierski [4, Theorem 1.1] have shown that if R is a right nonsingular ring, then every finitely generated nonsingular right R-module is torsionless precisely when Q_r is a left quotient ring of R (in the sense that $_RR$ is an essential submodule of $_RQ$).

Corollary 3.5. If $Q = Q_r$ and Q_ℓ is the maximal left quotient ring of R, then the following statements are equivalent.

(1) Every FG nonsingular M_R embeds in a free module.

(2) Q is a left quotient ring of R, Q_ℓ is right R-flat, and $R \to Q$ is a ring epimorphism.

(3) Q is a left quotient ring of R, Q_ℓ is right R-flat, and Q_R is (Π-) R-ML.

Proof. (1) ⇒ (2). Suppose that (1) holds. Then Q is a left quotient ring of R by [4, Theorem 1.1]. Since Q is regular, $0 = Z(_QQ) = Z(_RR)$. But if $Z(_RR) = 0$ then any regular left quotient ring of R is right R-flat if and only if R is left Z-coherent [3a, Theorem 2.1]. But $(Q_r)_R$ is flat since it is f-projective; hence Q_ℓ is right R-flat. By Theorem 3.4 (4), $R \to Q$ is a ring epimorphism.

(2) ⇒ (1). If (2) holds then $Z(_RR) = 0$ and Q_R is flat by the same arguments as in (1) ⇒ (2). The conclusion follows from Theorem 3.4.

(1) ⇒ (3). If (1) holds it is sufficient to show Q_R is R-ML, which follows from Theorem 3.4 (2).

(3) $\Rightarrow$ (1). If (3) holds then $Z({}_RR) = 0$ and Q_R is flat. Hence Q_R is f-projective (Proposition 2.7), so that (1) holds by Theorem 3.4 (2). □

The equivalence of (4) and (6) in the following theorem has also been shown by Cateforis [3, Theorem 2.3] and Goodearl [12, Theorem 2.5].

Theorem 3.6. The following statements are equivalent.

(1) Q is a left quotient ring of R and every FG torsionless M_R (${}_RX$) is projective.

(2) Q_R is f-projective and every FG torsionless M_R (${}_RX$) is projective.

(3) Q_R is f-projective, R is strongly left and right coherent, and $WGD(R) \leq 1$.

(4) Q_R is f-projective and R is right semihereditary.

(5) Every submodule of Q_R^I is f-projective.

(6) Every FG nonsingular M_R is projective.

Proof. The equivalence of (1) and (3)-(6) follows from Theorems 2.12 and 3.1, and Cateforis and Sandomierski [4, Theorem 1.1]. (1) and (3) together imply (2), and (2) implies (1) by Theorem 3.4 and Corollary 3.5. □

The conditions of Theorem 3.6 are not left-right symmetric [13, p. 156, exercise 15]. If the right (Goldie) dimension of R is finite, however, they are [13, p. 156, exercise 16], and the left and right maximal quotient rings of R coincide. In fact, if R is right finite-dimensional and Q_r is a left quotient ring of R, then finitely generated nonsingular (left and right) R-modules are projective and the left and right maximal quotient rings of R coincide (Proposition 3.8).

Lemma 3.7. The following statements are equivalent for R.

(1) Dim $(R_R) < \infty$ and Q_r is a left quotient ring of R.

(2) Dim $({}_R R) < \infty$ and Q_ℓ is a right quotient ring of R.

If either of these conditions holds then $Q_r = Q_\ell$.

Proof. (1) $\Rightarrow$ (2). Assume (1) holds. Then R is right semihereditary [25, Lemma 4.3], so that $Z(R_R) = 0$. Then Q_r is a semisimple ring [20, Theorem 1.6]. Since Q_r is a left quotient ring, $Z({}_R R) = 0$ (by the regularity of Q_r) and Q_r is a large left Q_r-submodule of Q_ℓ. But Q_ℓ is a semisimple left Q_r module, so we conclude that $Q_r = Q_\ell$. Then Q_ℓ is semisimple, so that dim $({}_R R) < \infty$ and we are done. □

Proposition 3.8. If dim $(R_R) < \infty$ and Q_r is a left quotient ring of R, then all FG nonsingular modules are projective and $Q_r = Q_\ell$.

Proof. If dim $R_R < \infty$ then R is right semihereditary [25, Lemma 4.3]; hence FG torsionless M_R are projective [21, Th. 2.6] and $Z(R_R) = 0$. The conclusion follows from the lemma and Theorems 3.6 and 2.11.

Proposition 3.9. Let $Q = Q_r$ be a left quotient ring of R.

(1) If R_R is Π-R-ML then Q_R is Π-R-ML.

(2) If R_R is Π-flat then Q_R is Π-flat.

(3) If R_R is Π-f-projective then Q_R is Π-f-projective.

Proof. (1) By Proposition 2.5 it is sufficient to show that FG nonsingular M_R embed in a FP module. Since Q is a left quotient ring, every FG nonsingular M_R is torsionless [4, Theorem 1.1] and hence embeds in a FP module by the assumption that R_R is Π-R-ML.

(2), (3) The proofs of these statements are essentially the same as the proof of (1), using Proposition 2.1 and Theorem 3.4, respectively. □

This paper constitutes a portion of the author's dissertation at Kent State University. The author is deeply indebted to her advisors, Professors F. L. Sandomierski and D. R. Turnidge, for their continuing advice and constant encouragement.

BIBLIOGRAPHY

[1] F. W. Anderson and K. R. Fuller, Rings and Categories of Modules. Graduate Texts in Mathematics 13, New York: Springer-Verlag, 1973.

[2] H. Cartan and S. Eilenberg, Homological Algebra. Princeton: Princeton University Press, 1956.

[3] V. C. Cateforis, On regular self-injective rings, Pac. J. Math., 30 (1969), 39-45.

[3a] ________, Flat regular quotient rings, Trans. Amer. Math. Soc., 138 (1969), 241-249.

[4] V. C. Cateforis and F. L. Sandomierski, On modules of singular submodule zero, Can. J. Math., 23 (1971), 345-354.

[5] S. U. Chase, Direct products of modules, Trans. Amer. Math. Soc., 97 (1960), 457-473.

[6] T. Cheatham and E. Enochs, Injective hulls of flat modules, Comm. in Alg., 8 (1980), 1989-1995.

[7] T. G. Clarke, On $\aleph_{-1}$-projective modules, Ph.D. Thesis, Kent State University, 1976.

[8] R. R. Colby, Rings which have flat injective modules, J. Algebra, 35 (1975), 239-252.

[9] R. R. Colby and E. A. Rutter, Jr., Π-flat and Π-projective modules, Arch. Math., 22 (1971), 246-251.

[10] J. S. Golan, Localization of Noncommutative Rings. Pure and Applied Mathematics 30. New York: Marcel Dekker, Inc., 1975.

[11] K. R. Goodearl, Distributing tensor product over direct product, Pacific J. Math., 43 (1972), 107-110.

[12] _______, Singular torsion and the splitting properties, Memoirs of the Amer. Math. Soc., 124 (1972).

[13] _______, Ring Theory: Nonsingular Rings and Modules. Pure and Applied Mathematics 33. New York-Basel: Marcel Dekker, Inc., 1976.

[14] M. F. Jones, Coherence and torsion theories, Ph.D. Thesis, Kent State University, 1978.

[15] M. F. Jones, f-Projectivity and flat epimorphisms, Commun. in Alg., 9 (1981), 1603-1616.

[16] M. F. Jones, Coherence relative to an hereditary torsion theory, Comm. in Alg., to appear (1982).

[17] D. Lazard, Autour de la Platitude, Bull. Soc. Math. France, 97 (1969), 81-128.

[18] H. Lenzing, Endlich präsentierbare Moduln, Arch. Math., 20 (1969), 262-266.

[19] N. Popescu and T. Spircu, Quelques observations sur les épimorphismes plats (à gauche) d'anneaux, J. Algebra, 16 (1970), 40-59.

[20] F. L. Sandomierski, Semisimple maximal quotient rings, Trans. Amer. Math. Soc., 128 (1967), 112-120.

[21] _______, Nonsingular rings, Proc. Amer. Math. Soc., 19 (1968), 225-230.

[22] D. Simson, ℵ-flat and ℵ- projective modules, Bull. Acad. Polon. Sci. Ser. Sci. Math. Astron. Phys., 20 (1972), 109-114.

[23] B. Stenström, Rings and Modules of Quotients. Lecture Notes in Mathematics 237. Berlin-Heidelberg-New York: Springer-Verlag, 1971.

[24] D. R. Turnidge, Torsion theories and semihereditary rings, Proc. Amer. Math. Soc., 24 (1970), 137-143.

[25] R. Warfield, Serial rings and finitely presented modules, J. Algebra 37 (1975), 187-222.

[26] T. Würfel, Kohärenz und lokalisierung, Seminar F. Kasch, B. Pareigis Algebra-Berichte, Nr. 14, 1973, Mathematisches Institut der Universität München.

CONSTRUCTION OF UNIVERSAL MATRIX LOCALIZATIONS

Peter Malcolmson
Department of Mathematics
Wayne State University
Detroit, Michigan 48202

Given a collection Σ of square matrices over a ring R, the <u>universal Σ-inverting homomorphism</u> $\lambda: R \to R_\Sigma$ is the universal homomorphism carrying the elements of Σ to invertible matrices. This has been considered by P.M. Cohn and others. It is generally constructed by generators and relations, which method gives little insight into (for example) the kernel of λ. In this article I propose another construction of $\lambda: R \to R_\Sigma$ under a mild closure condition on Σ. Some information about λ may be derived, depending on how matrices in Σ can be factored.

In the first part of the article we present the definitions and results, together with some explanatory material. The proofs are relegated to the second part.

The Statements

Let R be an associative ring with unit (which is preserved by ring homomorphisms). An <u>R-ring</u> will mean a ring homomorphism from R to some other such ring. These objects form a category with morphisms being ring homomorphisms which make the obvious triangular diagrams commutative.

For Σ a collection of square matrices over R, an R-ring $\phi: R \to S$ is said to be <u>Σ-inverting</u> if the image under ϕ of every element of Σ is invertible over S. A Σ-inverting R-ring is <u>universal</u> if it factors uniquely through any Σ-inverting R-ring. Such an object is unique up to (unique) isomorphism of R-rings.

These definitions are from Cohn ([1], Chap. 7) in which the universal Σ-inverting ring is constructed using generators and relations. Cohn discusses the conditions under which R_Σ is a local ring, leading to the definition of a "prime matrix ideal." The author has used so called "zigzag" methods to obtain similar results ([2]) and these methods will again be used in the present effort.

To describe this method, let us assume first that the collection Σ of square matrices satisfies the following two conditions: 1) the 1×1 identity matrix is in Σ, and 2) if A and B are in Σ and if C is of the appropriate size, then $\begin{pmatrix} A & C \\ 0 & B \end{pmatrix}$ is in Σ. Under such conditions Σ is called <u>multiplicative</u>.

When Σ is multiplicative, Cohn has shown that every element of R_Σ is an entry in the inverse of the image in R_Σ of some element of Σ. Thus every element of R_Σ is of the form $\lambda(f)\lambda(A)^{-1}\lambda(x)$, where $A \in \Sigma$ (say $n \times n$), f is $1 \times n$ and x is $n \times 1$, all over R. The basis of the zigzag method is to construct R_Σ as a set of equivalence classes of such triples (f, A, x). The equivalence class of (f, A, x) is thus to be interpreted as the element $fA^{-1}x$ of R_Σ, with addition and multiplication defined according to that interpretation.

To this end, assume Σ is multiplicative set of square matrices over R and let T_Σ consist of all triples (f, A, x), where $A \in \Sigma$ and where (letting A be $n \times n$) f is $1 \times n$ and x is $n \times 1$, both over R. We will say that "f is a row the size of A" to describe this sort of shape, and similarly for the "column" x. Other elements of T_Σ will be denoted by (g, B, y), (h, C, z), etc.

Define a relation $\sim$ among elements of T_Σ by $(f, A, x) \sim (g, B, y)$ if there exist $L, M, P, Q \in \Sigma$, rows j and u the sizes of L and P, respectively, and columns w and v the sizes of M and Q, respectively, such that

$$\begin{pmatrix} A & 0 & 0 & 0 & x \\ 0 & B & 0 & 0 & -y \\ 0 & 0 & L & 0 & 0 \\ 0 & 0 & 0 & M & w \\ \hline f & g & j & 0 & 0 \end{pmatrix} = \begin{pmatrix} P \\ \hline u \end{pmatrix} \begin{pmatrix} Q \mid v \end{pmatrix} .$$

Thus PQ is a block-diagonal matrix; we have written 0 for zero blocks, rows and columns as necessary.

Following our interpretation of (f, A, x) as $fA^{-1}x$, we can see why this might be the correct definition (though not why it is so complicated) as follows: If all elements of Σ are invertible, then

$$\begin{aligned} 0 = uv &= uQ(PQ)^{-1} Pv \\ &= fA^{-1}x - gB^{-1}y + jL^{-1}0 + 0M^{-1}w \\ &= fA^{-1}x - gB^{-1}y . \end{aligned}$$

Thus $fA^{-1}x$ should be the same as $gB^{-1}y$.

LEMMA 1. The relation $\sim$ is an equivalence relation.

Let R_Σ denote the set of equivalence classes $T_\Sigma/\sim$, and denote the equivalence class containing (f, A, x) by $(f/A\backslash x)$, reminding us of $fA^{-1}x$. Again following our interpretation, we are led to the appropriate definitions of operations in R_Σ (as in [2]). For $(f/A\backslash x)$, $(g/B\backslash y) \in R_\Sigma$, define

$$(f/A\backslash x) + (g/B\backslash y) = \left((f\ g) \Big/ \begin{pmatrix} A & 0 \\ 0 & B \end{pmatrix} \Big\backslash \begin{pmatrix} x \\ y \end{pmatrix} \right) ,$$

$$(f/A\backslash x) \cdot (g/B\backslash y) = \left((f\ 0) \Big/ \begin{pmatrix} A & -xg \\ 0 & B \end{pmatrix} \Big\backslash \begin{pmatrix} 0 \\ y \end{pmatrix} \right) ,$$

$$-(f/A\backslash x) = (f/A\backslash -x) .$$

Also define a map $\lambda : R \to R_\Sigma$ by $\lambda(r) = (1/1\backslash r)$. All these make sense because Σ is multiplicative.

THEOREM. The above definitions give R_Σ a well-defined structure of associative ring with unit. Further, the map $\lambda : R \to R_\Sigma$ is the universal Σ-inverting R-ring, and each element $(f/A\backslash x)$ of R_Σ satisfies

$$(f/A\backslash x) = \lambda(f)\,\lambda(A)^{-1}\,\lambda(x)\ .$$

Corollary. An element $r \in R$ is in the kernel of λ if and only if there exist $L, M, P, Q \in \Sigma$, rows j and u the sizes of L and P, respectively, and columns w and v the sizes of M and Q, respectively such that

$$\left(\begin{array}{cc|c} L & 0 & 0 \\ 0 & M & w \\ \hline j & 0 & r \end{array}\right) = \left(\begin{array}{c} P \\ \hline u \end{array}\right) \left(\begin{array}{c|c} Q & v \end{array}\right).$$

The Proofs.

The proofs that follow will primarily be complicated factorizations of block matrices, as suggested by the definition of the equivalence relation. To make these easier to read, zeros will be replaced by dots and the matrices corresponding to L and M in the definition of $\sim$ will be outlined. Thus the factorization in the definition would be written:

$$(*)\qquad \left(\begin{array}{cccc|c} A & \cdot & \cdot & \cdot & x \\ \cdot & B & \cdot & \cdot & -y \\ \cdot & \cdot & \boxed{L} & \cdot & \cdot \\ \cdot & \cdot & \cdot & \boxed{M} & w \\ \hline f & g & j & \cdot & \cdot \end{array}\right) = \left(\begin{array}{c} P \\ \hline u \end{array}\right) \left(\begin{array}{c|c} Q & v \end{array}\right).$$

We will also denote by I the identity matrix and by E_i the row (or column) block matrix which is zero in each block except for an identity matrix in the i-th block. The size and shape of these matrices will be indicated by context. For example, if $P = \begin{pmatrix} A & 0 \\ 0 & B \end{pmatrix}$ is a block matrix, then

$E_2 P = (0\ B)$ and $E_1 E_2 P = \begin{pmatrix} 0 & B \\ 0 & 0 \end{pmatrix}$.

As an example of the techniques we will use, let us show that if L or M is a "null matrix" (i.e. does not appear) in a factorization (*), then there is a similar one in which they do appear.

<u>Proposition 1</u>. If (f, A, x) , (g, B, y) in T_Σ are such that there is a factorization of any of these forms (with L, M, P, Q $\in \Sigma$):

(a) $$\left(\begin{array}{cc|c} A & \cdot & x \\ \cdot & B & -y \\ \hline f & g & \cdot \end{array}\right) = \left(\begin{array}{c} P \\ \hline u \end{array}\right) \left(\begin{array}{c|c} Q & v \end{array}\right) \;;$$

(b) $$\left(\begin{array}{ccc|c} A & \cdot & \cdot & x \\ \cdot & B & \cdot & -y \\ \cdot & \cdot & L & \cdot \\ \hline f & g & j & \cdot \end{array}\right) = \left(\begin{array}{c} P \\ \hline u \end{array}\right) \left(\begin{array}{c|c} Q & v \end{array}\right) \;; \cdot$$

(c) $$\left(\begin{array}{ccc|c} A & \cdot & \cdot & x \\ \cdot & B & \cdot & -y \\ \cdot & \cdot & M & w \\ \hline f & g & \cdot & \cdot \end{array}\right) = \left(\begin{array}{c} P \\ \hline u \end{array}\right) \left(\begin{array}{c|c} Q & v \end{array}\right) \;;$$

then there is a factorization of the form (*) .

<u>Proof</u>:

(a) $$\left(\begin{array}{cccc|c} A & \cdot & \cdot & \cdot & x \\ \cdot & B & \cdot & \cdot & -y \\ \cdot & \cdot & \boxed{1} & \cdot & \cdot \\ \cdot & \cdot & \cdot & \boxed{1} & 1 \\ \hline f & g & 1 & \cdot & \cdot \end{array}\right) = \left(\begin{array}{ccc} P & \cdot & \cdot \\ \cdot & 1 & \cdot \\ \cdot & \cdot & 1 \\ \hline u & 1 & \cdot \end{array}\right) \left(\begin{array}{ccc|c} Q & \cdot & \cdot & v \\ \cdot & 1 & \cdot & \cdot \\ \cdot & \cdot & 1 & 1 \end{array}\right) \;;$$

(b) $$\left(\begin{array}{cccc|c} A & \cdot & \cdot & \cdot & x \\ \cdot & B & \cdot & \cdot & -y \\ \cdot & \cdot & \boxed{L} & \cdot & \cdot \\ \cdot & \cdot & \cdot & \boxed{1} & 1 \\ \hline f & g & j & \cdot & \cdot \end{array}\right) = \left(\begin{array}{cc} P & \cdot \\ \cdot & 1 \\ \hline u & \cdot \end{array}\right) \left(\begin{array}{cc|c} Q & \cdot & v \\ \cdot & 1 & 1 \end{array}\right) \;;$$

(c)
$$\left(\begin{array}{cccc|c} A & \cdot & \cdot & \cdot & x \\ \cdot & B & \cdot & \cdot & -y \\ \cdot & \cdot & \boxed{M} & \cdot & \cdot \\ \cdot & \cdot & \cdot & M & w \\ \hline f & g & \cdot & \cdot & \cdot \end{array}\right) = \left(\begin{array}{cc} P & -E_3 \\ \cdot & I \\ \hline u & \cdot \end{array}\right) \left(\begin{array}{cc|c} Q & QE_3 & v \\ \cdot & M & w \end{array}\right) ,$$

where the last equation follows because

$$PQE_3 = E_3M = \begin{pmatrix} 0 \\ 0 \\ M \end{pmatrix} ,$$

$$uQE_3 = (f\ g\ 0)E_3 = 0 \text{ , etc.}$$

We remark that many of the proofs to follow could be simplified if Σ was assumed to be closed under multiplication by matrices invertible over R . We proceed for the more general Σ to improve the applicability of the results.

<u>Proof of Lemma 1</u>: For $(f, A, x) \in T_\Sigma$, the factorization below (with null L, M) proves $\sim$ is reflexive:

$$\left(\begin{array}{cc|c} A & \cdot & x \\ \cdot & A & -x \\ \hline f & f & \cdot \end{array}\right) = \left(\begin{array}{cc} A & -I \\ \cdot & I \\ \hline f & \cdot \end{array}\right) \left(\begin{array}{cc|c} I & I & 0 \\ \cdot & A & -x \end{array}\right) .$$

Now assume $(f, A, x) \sim (g, B, y)$ via tha factorization (*). Symmetry for $\sim$ is given by the following factorization:

$$\left(\begin{array}{cccccc|c} B & \cdot & \cdot & \cdot & \cdot & \cdot & y \\ \cdot & A & \cdot & \cdot & \cdot & \cdot & -x \\ \cdot & \cdot & B & \cdot & \cdot & \cdot & \cdot \\ \cdot & \cdot & \cdot & L & \cdot & \cdot & \cdot \\ \cdot & \cdot & \cdot & \cdot & M & \cdot & -w \\ \cdot & \cdot & \cdot & \cdot & \cdot & B & y \\ \hline g & f & \cdot & j & \cdot & \cdot & \cdot \end{array}\right) = \left(\begin{array}{ccc} B & E_2 P & \cdot \\ \cdot & P & -E_2 \\ \cdot & \cdot & I \\ \hline g & u & \cdot \end{array}\right) \left(\begin{array}{ccc|c} I & -E_2 & -I & \cdot \\ \cdot & Q & QE_2 & -v \\ \cdot & \cdot & B & y \end{array}\right) .$$

For transitivity, assume $(f, A, x) \sim (g, B, y)$ via (*) and also assume that $(g, B, y) \sim (h, C, z)$ via the following factorization (with $L', M', P', Q' \in \Sigma$):

$$\left(\begin{array}{cccc|c} B & \cdot & \cdot & \cdot & y \\ \cdot & C & \cdot & \cdot & -z \\ \cdot & \cdot & L & \cdot & \cdot \\ \cdot & \cdot & \cdot & M' & w' \\ \hline g & h & j' & \cdot & \cdot \end{array}\right) = \left(\frac{P'}{u'}\right)\left(Q' \,|\, v'\right) .$$

Then $(h, C, z) \sim (f, A, x)$ is justified by the factorization of the matrix

$$\left(\begin{array}{ccccccccccc|c} C & \cdot & \cdot & \cdot & \cdot & \cdot & \cdot & \cdot & \cdot & \cdot & \cdot & z \\ \cdot & A & \cdot & \cdot & \cdot & \cdot & \cdot & \cdot & \cdot & \cdot & \cdot & -x \\ \cdot & \cdot & B & \cdot & \cdot & \cdot & \cdot & \cdot & \cdot & \cdot & \cdot & \cdot \\ \cdot & \cdot & \cdot & L & \cdot & \cdot & \cdot & \cdot & \cdot & \cdot & \cdot & \cdot \\ \cdot & \cdot & \cdot & \cdot & M & \cdot & \cdot & \cdot & \cdot & \cdot & \cdot & \cdot \\ \cdot & \cdot & \cdot & \cdot & \cdot & L' & \cdot & \cdot & \cdot & \cdot & \cdot & \cdot \\ \cdot & \cdot & \cdot & \cdot & \cdot & \cdot & B & \cdot & \cdot & \cdot & \cdot & -y \\ \cdot & \cdot & \cdot & \cdot & \cdot & \cdot & \cdot & C & \cdot & \cdot & \cdot & z \\ \cdot & \cdot & \cdot & \cdot & \cdot & \cdot & \cdot & \cdot & L' & \cdot & \cdot & \cdot \\ \cdot & \cdot & \cdot & \cdot & \cdot & \cdot & \cdot & \cdot & \cdot & M' & \cdot & -w' \\ \cdot & \cdot & \cdot & \cdot & \cdot & \cdot & \cdot & \cdot & \cdot & \cdot & M & w \\ \hline h & f & g & j & \cdot & j' & \cdot & \cdot & \cdot & \cdot & \cdot & \cdot \end{array}\right)$$

into the product

$$\left(\begin{array}{ccccc} C & \cdot & \cdot & E_2P' & \cdot \\ \cdot & P & \cdot & E_2E_1P' & E_4 \\ \cdot & \cdot & L' & E_3P' & \cdot \\ \cdot & \cdot & \cdot & P' & \cdot \\ \cdot & \cdot & \cdot & \cdot & I \\ \hline h & u & j' & u' & \cdot \end{array}\right) \qquad \left(\begin{array}{ccccc|c} I & \cdot & \cdot & -E_2 & \cdot & \cdot \\ \cdot & Q & \cdot & -QE_2E_1 & -QE_4 & -v \\ \cdot & \cdot & I & -E_3 & \cdot & \cdot \\ \cdot & \cdot & \cdot & Q' & \cdot & -v' \\ \cdot & \cdot & \cdot & \cdot & M & w \end{array}\right) .$$

Proof of Theorem. First we prove that the operations are well-defined. Suppose that $(f, A, x) \sim (g, B, y)$ via (*); we wish to show first that $(f/A\backslash x) + (h/C\backslash z) = (g/B\backslash y) + (h/C\backslash z)$. According to the definition of addition above, this equation is justified by the factorization of

$$\left(\begin{array}{cccccccccc|c} A & \cdot & \cdot & \cdot & \cdot & \cdot & \cdot & \cdot & \cdot & \cdot & x \\ \cdot & C & \cdot & \cdot & \cdot & \cdot & \cdot & \cdot & \cdot & \cdot & z \\ \cdot & \cdot & B & \cdot & \cdot & \cdot & \cdot & \cdot & \cdot & \cdot & -y \\ \cdot & \cdot & \cdot & C & \cdot & \cdot & \cdot & \cdot & \cdot & \cdot & -z \\ \cdot & \cdot & \cdot & \cdot & \boxed{L} & \cdot & \cdot & \cdot & \cdot & \cdot & \cdot \\ \cdot & \cdot & \cdot & \cdot & \cdot & A & \cdot & \cdot & \cdot & \cdot & x \\ \cdot & \cdot & \cdot & \cdot & \cdot & \cdot & B & \cdot & \cdot & \cdot & -y \\ \cdot & \cdot & \cdot & \cdot & \cdot & \cdot & \cdot & L & \cdot & \cdot & \cdot \\ \cdot & \cdot & \cdot & \cdot & \cdot & \cdot & \cdot & \cdot & \cdot & M & w \\ \hline f & h & g & h & j & \cdot & \cdot & \cdot & \cdot & \cdot & \cdot \end{array}\right)$$

into the product

$$\left(\begin{array}{cccccc} A & \cdot & \cdot & \cdot & \cdot & E_1P \\ \cdot & C & \cdot & -I & \cdot & \cdot \\ \cdot & \cdot & B & \cdot & \cdot & E_2P \\ \cdot & \cdot & \cdot & I & \cdot & \cdot \\ \cdot & \cdot & \cdot & \cdot & L & E_3P \\ \cdot & \cdot & \cdot & \cdot & \cdot & P \\ \hline f & h & g & \cdot & j & u \end{array}\right) \qquad \left(\begin{array}{cccccc|c} I & \cdot & \cdot & \cdot & \cdot & -E_1 & \cdot \\ \cdot & I & \cdot & I & \cdot & \cdot & \cdot \\ \cdot & \cdot & I & \cdot & \cdot & -E_2 & \cdot \\ \cdot & \cdot & \cdot & C & \cdot & \cdot & -z \\ \cdot & \cdot & \cdot & \cdot & I & -E_3 & \cdot \\ \cdot & \cdot & \cdot & \cdot & \cdot & Q & v \end{array}\right) .$$

This shows addition on the left is well-defined. For addition on the right the factorization is similar; or we may refer to commutativity, below. Under the same equivalence (*), we get $-(f/A\backslash x) = -(g/B\backslash y)$ by simply changing the sign of v and w in (*).

To show multiplication is well-defined, again assume $(f/A\backslash x) = (g/B\backslash y)$ via (*). To show $(f/A\backslash x)\cdot(h/C\backslash z) = (g/B\backslash y)\cdot(h/C\backslash z)$ we use the factorization of

$$\left(\begin{array}{cccccccccc|c}
A & -xh & \cdot & \cdot & \cdot & \cdot & \cdot & \cdot & \cdot & \cdot & \cdot \\
\cdot & C & \cdot & \cdot & \cdot & \cdot & \cdot & \cdot & \cdot & \cdot & z \\
\cdot & \cdot & B & -yh & \cdot & \cdot & \cdot & \cdot & \cdot & \cdot & \cdot \\
\cdot & \cdot & \cdot & C & \cdot & \cdot & \cdot & \cdot & \cdot & \cdot & -z \\
\cdot & \cdot & \cdot & \cdot & \boxed{L} & \cdot & \cdot & \cdot & \cdot & \cdot & \cdot \\
\cdot & \cdot & \cdot & \cdot & \cdot & A & \cdot & \cdot & \cdot & -xh & \cdot \\
\cdot & \cdot & \cdot & \cdot & \cdot & \cdot & B & \cdot & \cdot & yh & \cdot \\
\cdot & \cdot & \cdot & \cdot & \cdot & \cdot & \cdot & L & \cdot & \cdot & \cdot \\
\cdot & \cdot & \cdot & \cdot & \cdot & \cdot & \cdot & \cdot & M & -wh & \cdot \\
\cdot & \cdot & \cdot & \cdot & \cdot & \cdot & \cdot & \cdot & \cdot & C & z \\
\hline
f & \cdot & g & \cdot & j & \cdot & \cdot & \cdot & \cdot & \cdot & \cdot
\end{array}\right)$$

into the product

$$\left(\begin{array}{ccccccc}
A & -xh & \cdot & \cdot & \cdot & E_1P & \cdot \\
\cdot & C & \cdot & \cdot & \cdot & \cdot & I \\
\cdot & \cdot & B & -yh & \cdot & E_2P & \cdot \\
\cdot & \cdot & \cdot & C & \cdot & \cdot & -I \\
\cdot & \cdot & \cdot & \cdot & L & E_3P & \cdot \\
\cdot & \cdot & \cdot & \cdot & \cdot & P & \cdot \\
\cdot & \cdot & \cdot & \cdot & \cdot & \cdot & I \\
\hline
f & \cdot & g & \cdot & j & u & \cdot
\end{array}\right)
\left(\begin{array}{ccccccc|c}
I & \cdot & \cdot & \cdot & \cdot & -E_1 & \cdot & \cdot \\
\cdot & I & \cdot & \cdot & \cdot & \cdot & -I & \cdot \\
\cdot & \cdot & I & \cdot & \cdot & -E_2 & \cdot & \cdot \\
\cdot & \cdot & \cdot & I & \cdot & \cdot & I & \cdot \\
\cdot & \cdot & \cdot & \cdot & I & -E_3 & \cdot & \cdot \\
\cdot & \cdot & \cdot & \cdot & \cdot & Q & -vh & \cdot \\
\cdot & \cdot & \cdot & \cdot & \cdot & \cdot & C & z
\end{array}\right) .$$

To show $(h/C\backslash z) \cdot (f/A\backslash x) = (h/C\backslash z) \cdot (g/B\backslash y)$ we use the factorization o

$$\left(\begin{array}{ccccccccccc|c}
C & -zf & \cdot & \cdot & \cdot & \cdot & \cdot & \cdot & \cdot & \cdot & \cdot & \cdot \\
\cdot & A & \cdot & \cdot & \cdot & \cdot & \cdot & \cdot & \cdot & \cdot & \cdot & x \\
\cdot & \cdot & C & -zg & \cdot & \cdot & \cdot & \cdot & \cdot & \cdot & \cdot & \cdot \\
\cdot & \cdot & \cdot & B & \cdot & \cdot & \cdot & \cdot & \cdot & \cdot & \cdot & -y \\
\cdot & \cdot & \cdot & \cdot & C & -zj & \cdot & \cdot & \cdot & \cdot & \cdot & \cdot \\
\cdot & \cdot & \cdot & \cdot & \cdot & L & \cdot & \cdot & \cdot & \cdot & \cdot & \cdot \\
\cdot & \cdot & \cdot & \cdot & \cdot & \cdot & C & zf & \cdot & \cdot & \cdot & \cdot \\
\cdot & \cdot & \cdot & \cdot & \cdot & \cdot & \cdot & A & \cdot & \cdot & \cdot & x \\
\cdot & \cdot & \cdot & \cdot & \cdot & \cdot & \cdot & \cdot & B & \cdot & \cdot & -y \\
\cdot & \cdot & \cdot & \cdot & \cdot & \cdot & \cdot & \cdot & \cdot & L & \cdot & 0 \\
\cdot & \cdot & \cdot & \cdot & \cdot & \cdot & \cdot & \cdot & \cdot & \cdot & M & w \\
\hline
h & \cdot & h & \cdot & h & \cdot & \cdot & \cdot & \cdot & \cdot & \cdot & \cdot
\end{array}\right)$$

into the product

$$\left(\begin{array}{cccccccc}
C & -zf & \cdot & \cdot & \cdot & \cdot & -I & \cdot \\
\cdot & A & \cdot & \cdot & \cdot & \cdot & \cdot & E_1P \\
\cdot & \cdot & C & -zg & -I & \cdot & I & -zu \\
\cdot & \cdot & \cdot & B & \cdot & \cdot & \cdot & E_2P \\
\cdot & \cdot & \cdot & \cdot & I & -zj & \cdot & \cdot \\
\cdot & \cdot & \cdot & \cdot & \cdot & L & \cdot & E_3P \\
\cdot & \cdot & \cdot & \cdot & \cdot & \cdot & I & \cdot \\
\cdot & \cdot & \cdot & \cdot & \cdot & \cdot & \cdot & P \\
\hline
h & \cdot & h & \cdot & \cdot & \cdot & \cdot & \cdot
\end{array}\right)
\left(\begin{array}{cccccccc|}
I & \cdot & \cdot & \cdot & \cdot & \cdot & I & \cdot \\
\cdot & I & \cdot & \cdot & \cdot & \cdot & \cdot & -E_1 \\
\cdot & \cdot & I & \cdot & I & \cdot & -I & \cdot \\
\cdot & \cdot & \cdot & I & \cdot & \cdot & \cdot & -E_2 \\
\cdot & \cdot & \cdot & \cdot & C & \cdot & \cdot & -zjE_3 \\
\cdot & \cdot & \cdot & \cdot & \cdot & I & \cdot & -E_3 \\
\cdot & \cdot & \cdot & \cdot & \cdot & \cdot & C & zfE_1 \\
\cdot & \cdot & \cdot & \cdot & \cdot & \cdot & \cdot & Q
\end{array}$$

The various identities for an associative ring with unit will be veri below with null L and M. We remark that the zero and unit are $\lambda(0)$ a $\lambda(1)$, respectively. For commutativity of addition, $(f/A\backslash x) + (g/B\backslash y) =$ $(g/B\backslash y) + (f/A\backslash x)$ by the following factorization:

$$\left(\begin{array}{cccc|c} A & \cdot & \cdot & \cdot & x \\ \cdot & B & \cdot & \cdot & y \\ \cdot & \cdot & B & \cdot & -y \\ \cdot & \cdot & \cdot & A & -x \\ \hline f & g & g & f & \cdot \end{array}\right) = \left(\begin{array}{cccc} A & \cdot & \cdot & -I \\ \cdot & B & -I & \cdot \\ \cdot & \cdot & I & \cdot \\ \cdot & \cdot & \cdot & I \\ \hline f & g & \cdot & \cdot \end{array}\right) \left(\begin{array}{cccc|c} I & \cdot & \cdot & I & \cdot \\ \cdot & I & I & \cdot & \cdot \\ \cdot & \cdot & B & \cdot & -y \\ \cdot & \cdot & \cdot & A & -x \end{array}\right) .$$

Associativity for both addition and multiplication follow from the reflexivity of $\sim$, since the two sides of the equation desired turn out to be identical.

To check that $\lambda(0)$ is an identity for addition requires $(f/A\backslash x) + (1/1\backslash 0) = (f/A\backslash x)$, which is verified by the following factorization:

$$\left(\begin{array}{ccc|c} A & \cdot & \cdot & x \\ \cdot & 1 & \cdot & \cdot \\ \cdot & \cdot & A & -x \\ \hline f & 1 & f & \cdot \end{array}\right) = \left(\begin{array}{ccc} A & \cdot & -I \\ \cdot & 1 & \cdot \\ \cdot & \cdot & I \\ \hline f & 1 & 0 \end{array}\right) \left(\begin{array}{ccc|c} I & \cdot & I & \cdot \\ \cdot & 1 & \cdot & \cdot \\ \cdot & \cdot & A & -x \end{array}\right) .$$

For $-(f/A\backslash x)$ to give an additive inverse requires $(f/A\backslash x) + (f/A\backslash -x) = (1/1\backslash 0)$, as verified by the following factorization:

$$\left(\begin{array}{ccc|c} A & \cdot & \cdot & x \\ \cdot & A & \cdot & -x \\ \cdot & \cdot & 1 & \cdot \\ \hline f & f & 1 & \cdot \end{array}\right) = \left(\begin{array}{ccc} A & -I & \cdot \\ \cdot & I & \cdot \\ \cdot & \cdot & 1 \\ \hline f & \cdot & 1 \end{array}\right) \left(\begin{array}{ccc|c} I & I & \cdot & \cdot \\ \cdot & A & \cdot & -x \\ \cdot & \cdot & 1 & \cdot \end{array}\right) .$$

Verification of distributivity requires larger matrices; to check $(f/A\backslash x) \cdot (h/C\backslash z) + (g/B\backslash y) \cdot (h/C\backslash z) = [(f/A\backslash x) + (g/B\backslash y)] \cdot (h/C\backslash z)$ requires the following factorization:

$$\left(\begin{array}{ccccccc|c}
A & -xh & \cdot & \cdot & \cdot & \cdot & \cdot & \cdot \\
\cdot & C & \cdot & \cdot & \cdot & \cdot & \cdot & z \\
\cdot & \cdot & B & -yh & \cdot & \cdot & \cdot & \cdot \\
\cdot & \cdot & \cdot & C & \cdot & \cdot & \cdot & z \\
\cdot & \cdot & \cdot & \cdot & A & \cdot & -xh & \cdot \\
\cdot & \cdot & \cdot & \cdot & \cdot & B & -yh & \cdot \\
\cdot & \cdot & \cdot & \cdot & \cdot & \cdot & C & -z \\
\hline
f & \cdot & g & \cdot & f & g & \cdot & \cdot
\end{array}\right) =$$

$$\left(\begin{array}{ccccccc}
A & -xh & \cdot & \cdot & -I & \cdot & \cdot \\
\cdot & C & \cdot & \cdot & \cdot & \cdot & -I \\
\cdot & \cdot & B & -yh & \cdot & -I & \cdot \\
\cdot & \cdot & \cdot & C & \cdot & \cdot & -I \\
\cdot & \cdot & \cdot & \cdot & I & \cdot & \cdot \\
\cdot & \cdot & \cdot & \cdot & \cdot & I & \cdot \\
\cdot & \cdot & \cdot & \cdot & \cdot & \cdot & I \\
\hline
f & \cdot & g & \cdot & \cdot & \cdot & \cdot
\end{array}\right)
\left(\begin{array}{ccccccc|c}
I & \cdot & \cdot & \cdot & I & \cdot & \cdot & \cdot \\
\cdot & I & \cdot & \cdot & \cdot & \cdot & I & \cdot \\
\cdot & \cdot & I & \cdot & \cdot & I & \cdot & \cdot \\
\cdot & \cdot & \cdot & I & \cdot & \cdot & I & \cdot \\
\cdot & \cdot & \cdot & \cdot & A & \cdot & -xh & \cdot \\
\cdot & \cdot & \cdot & \cdot & \cdot & B & -yh & \cdot \\
\cdot & \cdot & \cdot & \cdot & \cdot & \cdot & C & -z
\end{array}\right) .$$

For the reverse, $(h/C\backslash z) \cdot [(f/A\backslash x) + (g/B\backslash y)] = (h/C\backslash z) \cdot (f/A\backslash x) + (h/C\backslash z) \cdot (g/B\backslash y)$ requires the factorization:

$$\left(\begin{array}{ccccccc|c}
C & -zf & -zg & \cdot & \cdot & \cdot & \cdot & \cdot \\
\cdot & A & \cdot & \cdot & \cdot & \cdot & \cdot & x \\
\cdot & \cdot & B & \cdot & \cdot & \cdot & \cdot & y \\
\cdot & \cdot & \cdot & C & -zf & \cdot & \cdot & \cdot \\
\cdot & \cdot & \cdot & \cdot & A & \cdot & \cdot & -x \\
\cdot & \cdot & \cdot & \cdot & \cdot & C & -zg & \cdot \\
\cdot & \cdot & \cdot & \cdot & \cdot & \cdot & B & -y \\
\hline
h & \cdot & \cdot & h & \cdot & h & \cdot & \cdot
\end{array}\right)$$

$$= \left(\begin{array}{ccccccc} C & -zf & -zg & -I & \cdot & -I & \cdot \\ \cdot & A & \cdot & \cdot & -I & \cdot & \cdot \\ \cdot & \cdot & B & \cdot & \cdot & \cdot & -I \\ \cdot & \cdot & \cdot & I & \cdot & \cdot & \cdot \\ \cdot & \cdot & \cdot & \cdot & I & \cdot & \cdot \\ \cdot & \cdot & \cdot & \cdot & \cdot & I & \cdot \\ \cdot & \cdot & \cdot & \cdot & \cdot & \cdot & I \\ \hline h & \cdot & \cdot & \cdot & \cdot & \cdot & \cdot \end{array}\right) \left(\begin{array}{ccccccc|c} I & \cdot & \cdot & I & \cdot & I & \cdot & \cdot \\ \cdot & I & \cdot & \cdot & I & \cdot & \cdot & \cdot \\ \cdot & \cdot & I & \cdot & \cdot & \cdot & I & \cdot \\ \cdot & \cdot & \cdot & C & -zf & \cdot & \cdot & \cdot \\ \cdot & \cdot & \cdot & \cdot & A & \cdot & \cdot & -x \\ \cdot & \cdot & \cdot & \cdot & \cdot & C & -zg & \cdot \\ \cdot & \cdot & \cdot & \cdot & \cdot & \cdot & B & -y \end{array}\right) .$$

The proofs that λ is a homomorphism and that $\lambda(1)$ acts as a unit element are subsumed in the following:

Lemma 2. The following equations hold in R_Σ :

(i) $(f_1/A\backslash x) + (f_2/A\backslash x) = (f_1 + f_2/A\backslash x)$;

(i') $(f/A\backslash x_1) + (f/A\backslash x_2) = (f/A\backslash x_1 + x_2)$;

(ii) $\lambda(r) \cdot (f/A\backslash x) = (rf/A\backslash x)$;

(ii') $(f/A\backslash x) \cdot \lambda(r) = (f/A\backslash xr)$.

Proof: The statements are successively justified by the following factorizations:

(i) $$\left(\begin{array}{ccc|c} A & \cdot & \cdot & x \\ \cdot & A & \cdot & x \\ \cdot & \cdot & A & -x \\ \hline f_1 & f_2 & f_1+f_2 & \cdot \end{array}\right) = \left(\begin{array}{ccc} A & \cdot & -I \\ \cdot & A & -I \\ \cdot & \cdot & I \\ \hline f_1 & f_2 & \cdot \end{array}\right) \left(\begin{array}{ccc|c} I & \cdot & I & \cdot \\ \cdot & I & I & \cdot \\ \cdot & \cdot & A & -x \end{array}\right) ;$$

(i') $$\left(\begin{array}{ccc|c} A & \cdot & \cdot & x_1 \\ \cdot & A & \cdot & x_2 \\ \cdot & \cdot & A & -x_1-x_2 \\ \hline f & f & f & \cdot \end{array}\right) = \left(\begin{array}{ccc} A & -I & -I \\ \cdot & I & \cdot \\ \cdot & \cdot & I \\ \hline f & \cdot & \cdot \end{array}\right) \left(\begin{array}{ccc|c} I & I & I & \cdot \\ \cdot & A & \cdot & x_2 \\ \cdot & \cdot & A & -x_1-x_2 \end{array}\right) ;$$

(ii) $$\left(\begin{array}{ccc|c} 1 & -rf & \cdot & \cdot \\ \cdot & A & \cdot & x \\ \cdot & \cdot & A & -x \\ \hline 1 & \cdot & rf & \cdot \end{array}\right) = \left(\begin{array}{ccc} 1 & -rf & \cdot \\ \cdot & A & -I \\ \cdot & \cdot & I \\ \hline 1 & \cdot & \cdot \end{array}\right) \left(\begin{array}{ccc|c} 1 & \cdot & rf & \cdot \\ \cdot & I & I & \cdot \\ \cdot & \cdot & A & -x \end{array}\right) ;$$

(ii') $$\left(\begin{array}{ccc|c} A & -x & \cdot & \cdot \\ \cdot & 1 & \cdot & r \\ \cdot & \cdot & A & -xr \\ \hline f & \cdot & f & \cdot \end{array}\right) = \left(\begin{array}{ccc} A & -x & -I \\ \cdot & 1 & \cdot \\ \cdot & \cdot & I \\ \hline f & \cdot & \cdot \end{array}\right) \left(\begin{array}{ccc|c} I & \cdot & I & \cdot \\ \cdot & 1 & \cdot & r \\ \cdot & \cdot & A & -xr \end{array}\right) .$$

To show that the homomorphism $\lambda : R \to R_\Sigma$ is Σ-inverting, let A be an arbitrary matrix in Σ. We claim that the (i,j)-entry of $\lambda(A)^{-1}$ is $(E_i/A\backslash E_j)$, where here E_i and E_j denote a row and column respectivel To verify the claim on one side, we will show that $\sum_i \lambda(E_k A E_i)(E_i/A\backslash E_j) = \delta_{kj}$ the Kronecker delta. Using Lemma 2 successively, what we need to show is $(E_k A/A\backslash E_j) = \lambda(\delta_{kj})$. This is proved by the factorization

$$\left(\begin{array}{cc|c} A & \cdot & E_j \\ \cdot & 1 & -\delta_{kj} \\ \hline E_k A & 1 & \cdot \end{array}\right) = \left(\begin{array}{cc} I & \cdot \\ \cdot & 1 \\ \hline E_k & 1 \end{array}\right) \left(\begin{array}{cc|c} A & \cdot & E_j \\ \cdot & 1 & -\delta_{kj} \end{array}\right) .$$

A similar factorization proves that $(E_j/A\backslash E_j)$ works as a left inverse for $\lambda(A)$.

Further applications of Lemma 2 show that $(f/A\backslash x) = \lambda(f)\lambda(A)^{-1}\lambda(x)$. Now given a Σ-inverting R-ring $\varphi : R \to S$, we may define $\varphi^+ : R_\Sigma \to S$ by $\varphi^+(f/A\backslash x) = \varphi(f)\varphi(A)^{-1}\varphi(x)$. This is well-defined by the computation pre- ceding the statement of Lemma 1, and it is easy to check that φ^+ is a homo morphism satisfying $\varphi^+\lambda = \varphi$. Furthermore, φ^+ is the unique such homomorphism, since the inverse of $\varphi(A)$ is uniquely determined by $\varphi(A)$. Thus $\lambda : R \to R_\Sigma$ is the universal Σ-inverting R-ring.

<u>Proof of Corollary</u>: If $\lambda(r) = 0$, then there is a factorization as follows:

$$\left(\begin{array}{cccc|c} 1 & \cdot & \cdot & \cdot & r \\ \cdot & 1 & \cdot & \cdot & \cdot \\ \cdot & \cdot & L' & \cdot & \cdot \\ \cdot & \cdot & \cdot & M' & w' \\ \hline 1 & 1 & j' & \cdot & \cdot \end{array}\right) = \left(\begin{array}{c} P' \\ \hline u' \end{array}\right) \left(\begin{array}{c|c} Q' & v' \end{array}\right) ,$$

where $L', M', P', Q' \in \Sigma$, etc. Then the following factorization

$$\left(\begin{array}{ccc|cc|c} 1 & \cdot & \cdot & \cdot & \cdot & \cdot \\ \cdot & 1 & \cdot & \cdot & \cdot & \cdot \\ \cdot & \cdot & L' & \cdot & \cdot & \cdot \\ \cdot & \cdot & \cdot & M' & \cdot & -w' \\ \cdot & \cdot & \cdot & \cdot & 1 & r \\ \hline 1 & 1 & j' & \cdot & \cdot & r \end{array}\right) = \left(\begin{array}{cc} P' & E_1 \\ \cdot & 1 \\ \hline u' & 1 \end{array}\right) \left(\begin{array}{cc|c} Q' & -Q'E_1 & -v' \\ \cdot & 1 & r \end{array}\right)$$

allows us to put $j = (1\ 1\ j')$, etc. Conversely if there is a factorization as in the Corollary then $\lambda(r) = 0$ follows from the factorization:

$$\left(\begin{array}{cccc|c} 1 & \cdot & \cdot & \cdot & \cdot \\ \cdot & 1 & \cdot & \cdot & -r \\ \cdot & \cdot & L & \cdot & \cdot \\ \cdot & \cdot & \cdot & M & w \\ \hline 1 & 1 & j & \cdot & \cdot \end{array}\right) = \left(\begin{array}{ccc} 1 & \cdot & \cdot \\ \cdot & 1 & \cdot \\ \cdot & \cdot & P \\ \hline 1 & 1 & u \end{array}\right) \left(\begin{array}{ccc|c} 1 & \cdot & \cdot & \cdot \\ \cdot & 1 & \cdot & -r \\ \cdot & \cdot & Q & v \end{array}\right) .$$

References

[1] P.M. Cohn, Free Rings and their Relations, Academic Press, London, 1971.

[2] P. Malcolmson, "A Prime Matrix Ideal Yields a Skew Field", J. London Math. Soc. (2), 18(1978), 221-233.

ARITHMETICAL ZARISKI CENTRAL RINGS

Erna Nauwelaerts
L.U.C. Hasselt, Belgium

Jan Van Geel
University of Antwerp, U.I.A., Belgium

Introduction.

In (6) F. Van Oystaeyen introduced Zariski central rings (ZCR). ZCR's arise in a natural way as a class of rings having nice properties with respect to symmetric localisation at prime ideals, e.g. symmetric localisation at all prime ideals is actually a central localisation (cf. Proposition 2). As a consequence one finds that within the class of ZCR's many subclasses of rings may be characterised in a local-global way, cf. (2), (3), (6).
Since the ideals of a ZCR are strongly linked to the ideals of the center, one may expect that the class of ZCR's also yields a natural framework to study arithmetical structures on the sets of ideals. In (6) F. Van Oystaeyen already obtained some results of an arithmetical nature for ZCR's. The theory of Zariski central Asano orders, Dedekind rings and Hereditary orders was worked out by E. Nauwelaerts and F. Van Oystaeyen in (2), (3).
In this paper we focus on ZCR's (not necessarily orders) for which the ideals form an abelian group, these are called <u>arithmetical ZCR.</u> It turns out that these rings closely generalise the concept of commutative Dedekind rings (Theorem 4). In section 2 we show that it is possible to associate to prime ideals of an arithmetical ZCR R certain value functions (arithmetical pseudovaluations, a.p.v.) on the symmetric total ring of quotients, $Q_{sym}(R)$. These a.p.v.'s are in one to one correspondence with primes in $Q_{sym}(R)$. (The concept of primes was studied by J. Van Geel in (4), (5), and seems to be a useful generalisation of valuation theory to the noncommutative case).
It is mentioned that the set of a.p.v. (primes) in $Q_{sym}(R)$ yields an alternative description of the ideal theory of R, however this has

not been worked out in full detail here.
The fact that primes in $Q_{sym}(R)$ can also obtained by symmetric localisation of R at prime ideals (corallary 11), completes the analogy with the theory of commutative Dedekind rings.
In the last section we consider a few examples which provide some insight in the relationship between the class of arithmetical rings, Asano orders and Dedekind rings.
We owe much to our PhD-advisor, F. Van Oystaeyen. Dedicating this paper to him for his 33th anniversary will probably not clear off our debt.

1. Zariski extensions.

Let A be a ring, Spec A its prime spectrum endowed with the Zariski topology.
A ring homomorphism f: A $\to$ B is said to be an extension if $B=f(A)Z_B(A)$ where $Z_B(A)=\{b \in B|\ bf(a)=f(a)b,$ for all $a \in A\}$, cf. (2). If f: A $\to$ B is an extension, then $f^{-1}(P)$ is in Spec A for any P in Spec B, and the map ϕ: Spec B $\to$ Spec A given by $\phi(P)=f^{-1}(P)$ is continuous. Furthermore $f(H)B=Bf(H)$ for any ideal H of A. A monomorphic extension f:A $\to$ B is said to be a Zariski extension, cf. (2), if there exist nonempty Zariski open sets Y(I) (=$\{P \in$ Spec B$|$ I not included in P$\}$), X(J) subset of Spec A satisfying:
The restriction of "P $\to$ $f^{-1}(P)$" yields a homeomorphism between Y(I) and X(J), an open subset Y(H) of Y(I) corresponds to an open subset X(H') of X(J) with H' subset of $f^{-1}(H)$.
If f: A $\to$ B is a Zariski extension, then B is called a Zariski A-algebra. A global Zariski extension f: A $\to$ B is a Zariski extension such that the above condition is satisfied for Y(I)=Spec B and some open set of Spec A.
Let A $\hookrightarrow$ B be an extension, then ϕ: Spec B $\to$ Spec A is given by $\phi(P)= P\cap A$. For non empty open sets Y(I), X(J) the ZE. property is equivalent to: $\phi(Y(I))=X(J)$ and rad H=rad $B(H\cap A)$ for every ideal H in rad I.
Examples.
1. Clearly a simple ring and any semisimple (Artinian) ring are global Zariski extensions of their center.
2. Let R be any ring, the extension R $\to$ $M_n(R)$ is a global Zariski extension.
3. Any Azumaya algebra is a global Zariski extension of its center.
The following results on (semiprime) Zariski algebras will be used

further on.

Proposition 1. *Let B be a Zariski A-algebra described on the open sets* Y(I) *and* X(J). *Then:*

(1) *If* B *is a semiprime ring, then a non-trivial ideal of* B, *which is contained in* rad I, *intersects* A *non-trivially.*

(2) *If* B *is a prime ring, then every non-trivial ideal of* B *intersects* A *non-trivially.*

(3) *If* B *is semiprime and* A *is a simple ring, then* B *is a simple ring too.*

Proof: cf. (3)

Recall some definitions and facts concerning (symmetric) localisation. A kernel functor κ on R-mod is a left exact subfunctor of the identity in R-mod. Such a kernel functor κ is idempotent if, for any M in R-mod: $\kappa(M/_{\kappa(M)})=0$. There is a one-to-one correspondence between kernel functors κ and Gabriel filters $\mathcal{L}(\kappa)=\{L$ left ideal of R, $\kappa(R/L)=R/L\}$, making R into a left linearly topological ring. A kernel functor is said to be bilateral if $\mathcal{L}(\kappa)$ allows a cofinal set of ideals.
A symmetric kernel functor is a kernel functor which is idempotent and bilateral. To a prime ideal P of R we associate a kernel functor κ_{R-P}, given by its filter $\mathcal{L}(R-P)=\{L$ left ideal, $L\supset I$, I an ideal of R such that $I\not\subset P\}$. If R is left Noetherian then κ_{R-P} is idempotent.
All concepts introduced thus far have right analogues; if we want to distinguish between the kernel functors on R-mod or mod-R associated to the prime ideal P of R, we will write κ^{l}_{R-P} or κ^{r}_{R-P} resp., otherwise κ_{R-P} always stands for the kernel functor on R-mod associated to P.
To an idempotent kernel functor κ on R-mod there corresponds a left exact localisation functor Q_κ on R-mod. We say that κ has property (T) if Q_κ is right exact and commutes with direct sums, or equivalently, if $Q_\kappa(M)\cong Q_\kappa(R)\otimes_R M$ for all M in R-mod.
In case R is left Noetherian, κ has property (T) if and only if $\mathcal{L}(\kappa)$ has a basis consisting of κ-projective left ideals. (For more details cf. (1)).
The relation between kernel functors in B-mod and A-mod for a Zariski A-algebra B has been studied by F. Van Oystaeyen in (6), and by E. Nauwelaerts in (2).
Recall:

Proposition 2. *Let* $A\to B$ *be a Zariski extension,* Y(I), X(J) *the Zariski open sets. If* P *is in* Y(I) *then* $P\cap A=p$ *is in* X(J) *and the following properties hold:*

(1) M *in* B-mod *is* κ_{B-P}- *torsion if and only if* M *is* κ_{A-p}- *torsion as an* A-*module. Moreover* κ_{B-P} *is an idempotent kernel functor on* B-mod.
(2) $Q_{B-P}(M) \cong Q_{A-p}(M)$ *for every* M *in* B-mod,*and* $Q_{B-P}(B) \cong Q_{A-p}(A)$ *as rings.*
(3) κ_{B-P} *has property* (T), *and ideals of* B *extend to ideals of* $Q_{B-P}(B)$.
(4) $Q_{B-P}(B)$ *is a global Zariski* $Q_{A-p}(A)$-*algebra.*
Proof: cf. (3), (6).

Corollary 3. *Let* R *be a global Zariski extension of its center* C. *Then for all* P *in* Spec R *the following properties hold:*
(1) κ_{R-P} *is a central kernel functor; i.e. let* M be in R-mod *then* κ_{R-P} *viewed as* C-*module is equal to* $\kappa_{C-p}(M)$. *Moreover* κ_{R-P} *is an idempotent kernel functor on* R-mod.
(2) $Q^{l}_{R-P}(R)$, $Q^{r}_{R-P}(R)$ *and* $Q_{C-p}(R)$ *are isomorphic rings.*
(3) *Ideals of* R *extend to ideals of* $Q_{R-P}(R)$.
(4) $Q_{R-P}(R)$ *is a global Zariski extension of* $Q_{C-p}(C)$.

In view of the above propositions we have: If R is a prime ring which is a global Zariski extension of its center C, then $Q_{sym}(R)$ $(=Q_{R-0}^{\cdot}(R))$ is a simple ring containing R. Every non zero element c of C, is invertible in $Q_{sym}(R)$, and an element of $Q_{sym}(R)$ has the form $c^{-1}r=rc^{-1}$ for suitable c in C and r in R. Note that $Q_{sym}(R)$ need not to be a ring of fractions (i.e. regular elements need not be invertible.).
We define a fractional ideal of R in $Q_{sym}(R)$ to be a non-zero left and right R-module I of $Q_{sym}(R)$ such that $cI\subset R$ for some $c(\neq 0)$ in C. We say that the fractional ideal I is invertible if there exists a fractional ideal I' of R such that II'=I'I=R, we then write $I'=I^{-1}$. Observe also that if J is a non-zero left and right R-submodule of $Q_{sym}(R)$ such that $qJq'\subset R$ for some non-zero elements q,q' in $Q_{sym}(R)$, then $rJr'\subset R$ for suitable non-zero elements r,r' in R, whence $RrrJr'r\subset R$ and thus $cJ\subset R$ for some $c(\neq 0)$ of C; using Propositions 1 and 2. Moreover I contains a non-zero element of C. The next result yields conditions on a prime ring R which is a global Zariski extension of its center, in order to make it into an Asano order.

Theorem 4. *Let* R *be a prime ring which is a global Zariski extensio*
its center C. *Then the following conditions on* R *are equivalent:*
(1) *Every non-trivial ideal of* R *is the product of maximal ideals.*
(2) *Every ideal of* R *is the product of prime ideals, and hte produ*
ideals is commutative.
(3) *The fractional ideals of* R *in* $Q_{sym}(R)$ *form an abelian group*
multiplication.
(4) R *is not properly contained in any subring* S *of* $Q_{sym}(R)$ *suc*

$cS \subset R$ for some $c(\neq 0)$ in C, non-trivial prime ideals of R are maximal, and R satisfies the ascending chain condition for ideals.

(5) *The non-zero maximal ideals of R are invertible in $Q_{sym}(R)$, and R satisfies the ascending chain conditions for ideals.*

<u>Proof</u>: We may assume that R is not a simple ring.

(1)→(2) We only have to prove that the product of maximal ideals commutes.

Let M,N be two maximal ideals of R, put $M'=R(M \cap C)$ and $N'=R(N \cap C)$. We claim that M=N or $M'+N'=R$. Suppose $M'+N'$ is contained in some maximal ideal Q of R. Since R is a global Zariski C-algebra, $H \cap C \cap Q \cap C$ implies $H \subset Q$ for any ideal H of R. It follows that M and N are contained in Q so M=N or Q=R, which yields $M'+N'=R$. In case M=N there is nothing to prove. In the other case $M \cap N = R(M \cap N) = (M'+N')(M \cap N) \subset M'N+N'M$, but $N'M=MN'$ so $M \cap N \subset MN$; similary $M \cap N \subset NM$. Therefore $MN = M \cap N = NM$.

(2)→(3) In view of the hypotheses it is enough to show that non-trivial prime ideals P are invertible. Since R is a prime, Zariski C-algebra, P intersects the center non-trivially. Choose $a(\neq 0)$ in $P \cap C$; $Pa \subset P$. Then a has an inverse a^{-1} in $Q_{sym}(R)$, and $Ra^{-1}=a^{-1}R$ is a fractional ideal of R in $Q_{sym}(R)$ such that $RaRa^{-1}=R=Ra^{-1}Ra$. Now, by hypothesis Ra is a (commutative) product of prime ideals. One of these must be contained in P, say Q. Clearly since Ra is invertible Q must be invertible. Analoguous as in the case of commutative Dedekind rings one shows that Q is a maximal ideal, whence Q=P, yielding that P is invertible.

(3)→(4) If I is a non-zero ideal of R, then it is invertible so there exists an I^{-1} such that $I^{-1}I=R$. Then $1=\Sigma q_i a_i$ for some q_i in I^{-1} and a_i in I, and thus $a=\Sigma a q_i a_i$ for any a in I. As aq_i is in R it follows that every ideal of R satisfies the ascending chain condition for ideals. Next, let S be a subring of $Q_{sym}(R)$ containing R such that $cS \subset R$ for ome $c(\neq 0)$ in R. Then cS is a non-zero ideal of R, and thus, by assumpion, it is invertible. So $S=RS=(cS)^{-1}cS=(cS)^{-1}cS=R$. Finally, let P be non-trivial prime ideal of R and let I be an ideal of R such that I. By assumption I is invertible. Then $I^{-1}P \subset I^{-1}I=R$. Clearly $P \subset I^{-1}$ versely, if r is in $I^{-1}P$ then $Ir \subset P$, and thus r is an element of P. since P is also ivertible, $P=I^{-1}P$ implies I=R. Therefore P is a mal ideal of R.

5) First we prove that every ideal I of R contains a finite pro- of prime ideals, each of which contains I. If I is not a prime of R, then there exists ideals A and B of R such that $A \not\subset B$, d $AB \subset I$. Then $I \subsetneq I+A$, $I \subsetneq I+B$ and (I+A)(I+B) I. If either I+A is not a prime ideal, then we repeat the former construction. that any ascending chain of ideals of R must terminate comple-

tes the proof.
Let now M be a maximal ideal of R. First we note that M contains a non-zero element c of the center. Set $M'=\{q \in Q_{sym}(R) \mid Mq \subset R\}$. Evidently, MM' is an ideal of R containing M, hence either $MM'=M$ or $MM'=R$. We suppose that $MM'=M$. Then M' is a subring of $Q_{sym}(R)$ containing R, and $cM' \subset R$. By assumption, $M'=R$ follows. On the other hand, there exist prime ideals P_i, $i=1,\dots,n$, containing Rc such that $P_1 \dots P_n$ Rc, and we choose n minimal. One of these prime ideals must be contained in M. By hypothesis, every non-trivial prime ideal of R is maximal, and the product of maximal ideals is commutative (cf. proof of (1)→(2)). So we may assume that $P_1=M$. Now $P_2 \dots P_n$ is not contained in Rc. Let a be in $P_2 \dots P_n$ and a not in Rc, c has an inverse c^{-1} in $Q_{sym}(R)$. Then ac^{-1} is not in R and Mac^{-1} is contained in R, that is ac^{-1} is in M'; contradicting $M'=R$. Therefore $MM'=R$. In the same way one shows that $M''M=R$, where $M''=\{q \in Q_{sym}(R) \mid qM \subset R\}$. Moreover, $M'=M''$ and M' is a fractional ideal of R in $Q_{sym}(R)$.
(5)→(1) Note that $\bigcap_{i=1}^{\infty} M^i=0$ for any maximal ideal M of R, since otherwise $\bigcap M^i$ would contain a non-zero element of the center. M is invertible by hypothesis, and $M^{-i}(M^i \cap Rc)$ is an ideal of R for each integer $i \geqq 1$. Since R satisfies the ascending chain condition for ideals, there exists an integer $k \geqq 1$ such that; $M^{-k}(M^k \cap Rc) \subset \sum_{i=1}^{k-1} M^{-i}(M^i \cap Rc)$. Hence $M^k \cap Rc \subset \sum_{i=1}^{k-1} M^{k-i}(M^i \cap Rc) \subset MRc$. It follows that $Rc \subset M^k \cap Rc \subset MRc$. But the $Rc=Mc$, and thus $R=M$ which is false. Let now I be a non-trivial ideal M_1 of R. M_1 is invertible so $I \subset M_1^{-1} I \subset M_1^{-1} M_1 = R$. Moreover $I \neq M_1^{-1} I$. For if not, then $I=M_1 I=M_1^2 I=\dots$, whence I $\bigcap_{i=1}^{\infty} M^i=0$. If $M_1^{-1} I \neq R$, then the ideal M_1^{-1} is contained in some maximal ideal M_2, and $M_1^{-1} I \subsetneqq M_2^{-1} M_1^{-1} I \subset R$. Since R satisfies the ascending chain condition for ideals, it follows that $M_n^{-1} \dots M_2^{-1} M_1^{-1} I = R$ for some $n \geqq 1$, where the M_i are maximal ideals of R. So $I=M_1 \dots M_n$ as desired. |

Definition 5. Rings satisfying the properties of Theorem will be called arithmetical (ZC) rings.

We now assume that R is a left order in a quotient ring Q. We may adapt the proof of the theorem in order to obtain:

Corollary 6. *Let* R *be a left order in a quotient ring* Q, *and assume that* R *is a prime ring which is a global Zariski extension of its center* C. *Then the following conditions on* R *are equivalent:*
(1) R *is an Asano left order in* Q.
(2) *Every ideal of* R *is a product of prime ideals and the product of ideals of* R *is commutative.*

(3) *Every non-trivial ideal of R is a product of maximal ideals.*
(4) *R is a maximal left order in Q, non-trivial prime ideals of R are maximal, and R satisfies the ascending chain condition for ideals.*
(5) *The (non-zero) maximal ideals of R are invertible in Q, and R satisfies the ascending chain condition for ideals.*

2. Primes and Arithmetical pseudo valuations.

Let R be a ring with unit.
A pair (P,R') is called a prime in R iff it satisfies the following properties:
(1) R' is a subring of R
(2) P is a prime ideal in R'
(3) For all x,y in R; $xR'y \subset P$ yields x in P or y in P.
If (P,R') is a prime in R then (P,R^P), with $R^P=\{r \in R \mid rP\subset P \text{ and } Pr\subset P\}$ is also a prime in R. Note that R^P is maximal with this property.
The following result implies that primes in rings may be extended to primes in overrings.

Proposition 7. *A pair (P,R^P) in R is maximal with respect to the following relation:*
$(P_1,R^P1)<(P_2,R^P2) \leftrightarrow R^P1\subset R^P2$ *and* $P_2\cap R^P1 = P_1$
is a prime in R, called dominating prime in R.
Proof: cf.(5).

Examples.
1) Primes are natural generalisations of valuation rings in fields: If R=K is a field then (P,K^P) is a prime in K if and only if K^P is a valuation ring in K and P is its maximal ideal.
Note thet the above proposition applied in the case of fields yields the Krull-Chevalley extension theorem for valuations.
2) Maximal orders in central simple algebras A, lying over discrete valuation rings, yield primes in A, cf.(5).
In the sequel R is an arithmetical (ZC) ring (Definition 5), $Q=Q_{sym}(R)$ as in section 1, $\mathcal{F}$ the group of fractional ideals of R.
An arithmetical pseudo valuation (a.p.v.) v on $\mathcal{F}$ is a function $v:\mathcal{F} \to \Gamma\cup\{\infty\}$, where Γ is a totally ordered group, such that:
1) $v(IJ) = v(I)+v(J)$ for all I,J in $\mathcal{F}$
2) $v(I+J) \geq \min (v(I),v(J))$ for all I,J in $\mathcal{F}$
3) $v(R) = 0$, $v(0) = \infty$
4) If $I\subset J$ then $v(I) \geq v(J)$ for all I,J in $\mathcal{F}$

Remark. 1) In view of 4) property 3) may be strengthened to $v(I+J)=\min(v(I),v(J))$.
2) Note that this terminology is compatible with the existing concept of pseudovaluations on rings, since if $v:\mathcal{F}\to\Gamma\cup\{\infty\}$ is a pseudo valuation on fractional ideals then $v^*:Q\to\Gamma\cup\{\infty\};x\to v(RxR)$ defines a pseudo valuation on Q, such that $v^*(r)\geq 0$ for all r in R. The restriction of an a.p.v. to the center of Q defines a valuation.
3) Two a.p.v. 's are called equivalent iff $v_1(RxR)>0 \leftrightarrow v_2(RxR)>0$, for all x in R.
We have the following relation between a.p.v. on $\mathcal{F}$ and primes in Q:

Proposition 8. *Let* v *be an a.p.v. on* $\mathcal{F}$ *then* $P=\{x\in Q|\ v(RxR)>0\}$ *yields a prime* (P,Q^P) *in* Q.
Conversely if (P,Q^P) *is a prime in* Q, *such that* R Q^P, *then there exists an a.p.v. on* $\mathcal{F}$, *say* v, *such that* $P=\{x\in Q|\ v(RxR)>0\}$.
Proof: Clearly P is a multiplicatively closed additive subgroup of R_+, by definition of v. This implies that P is an ideal in Q^P. If x,y are elements of Q such that $xQ^Py\subset P$, then since $R\subset Q^P$, $xRy\subset P$. This implies $v(RxRyR)>0$ but since v is an a.p.v., $0<v(RxRyR)=v(RxR)+v(RyR)$ so either $v(RxR)>0$ or $v(RyR)>0$, i.e. x in P or y in P. Therefore (P,Q^P) is a prime in Q.
Conversely, given a prime (P,Q^P) in Q, with $R\subset Q^P$, define: $v(I)=\{x\in Q|\ xI\subset P\}$, for all fractional ideals I of R in Q. Let $\Gamma=\{v(I)|I\in\mathcal{F}\}$, then Γ is totally ordered by $v(I)\leq v(J)$ iff $v(I)\subset v(J)$. To show that this defines a total order on Γ, suppose for some I,J in $\mathcal{F}$ that $v(J)\not\subset v(I)$. Then there exists an element x in Q such that $xI\subset P$ and $xJ\not\subset P$, and also an element y in Q such that $yJ\subset P$ and $yI\not\subset P$. This yields $RxRJQ^PRyRI\not\subset P$, since P is a prime in Q, so $RxRJRzRyRI\not\subset P$ for some z in Q. But the fractional ideals of R commute so $RxRJRzRyRI=RyRJRzRxRI\not\subset P$, a contradiction.
Claim: $v(I)+v(J)\overset{D}{=}v(IJ)$ is a well defined addition on Γ, for which $v(R)$ is a unit element. This makes Γ into an ordered group. If $v(I)=v(I')$ and $v(J)=v(J')$ then for x in $v(IJ)$, $RxRIJ\subset P$, so $RxRI\subset v(J)=v(J')$, hence $RxRIJ'=RxRJ'I\subset P$, finally since $v(I)=v(I')$, $RxRJ'I=RxRI'J'\subset P$ follows, i.e. x is in $v(I'J')$.
The fact that $v(R)$ is a unit is obvious.
Now $v(I)\leq v(J)$ yields $v(I)+v(H)\leq v(J)+v(H)$ for all $v(H)$ in Γ. Indeed, if x is in $v(IH)$ then $xIH\subset P$ so $xHI\subset P$, i.e. $xH\subset v(I)\subset v(J)$ yielding $xHJ=xJH\subset P$, so x is in $v(JH)$.
Properties 1) - 4) of pseudo valuations follow directly from the definition of v, and $P=\{x|v(RxR)>0\}$.

Corollary 9. *If* (P,Q^P) *is a prime in* Q *with* $R\subset Q^P$, *then* $P\cap R$ *is a prime ideal in* R.

Remark. 1) The a.p.v. associated with a given prime (P,Q^P), defined as in Proposition 8., i.e. $v(I)=\{x|xI \quad P\}$, is denoted by v_P. And the prime associated with a given a.p.v. v, is denoted with (P_v,Q^{P_v}).
It is easy to see that with these notations, for a given a.p.v. v: v_{P_v} is equivalent to v.
2) Let (P,Q^P) be a prime in Q, $R\subset Q^P$, v_P its associated a.p.v.. For every fractional ideal I, the set $\{v_P(RxR)|x \in I\}$ has an infimum and $v_P(I)=\inf\{v_P(RxR)|x \in I\}$.

Proposition 10. *Let* (P,Q^P) *be a prime in* Q, $R\subset Q^P$, v_P *the associated valuation . Then*
1) $Q^P=\{x \in Q|v(RxR)\geq 0\}$
2) (P,Q^P) *is a dominating prime.*

Proof: We claim that for every fractional ideal I of R there is an element x in Q such that $xI\subset Q^P$ and $xI \not\subset P$.
For any fractional ideal I, $v(I)$ has an inverse; i.e. there exists a fractional ideal I such that $v(IJ)=v(R)=0$. By definition of v it follows that $IJ \not\subset P$, on the other hand since $P\subset v(R)$, $P\subset v(IJ)$ yields $IJ\subset Q^P$. Therefore there must be an element x in J with $xI \not\subset P$ and $xI\subset Q^P$.
Now obviously: $\{x \in Q|v(RxR)\geq 0\}\subset Q^P$. Take y in Q^P, then by the above property there exists a z in Q such that $v(RyR)+v(RzR)=v(RyRRzR)=0$.
And if $v(RyR)<0$ then $v(RzR)>0$ follows. But since y is in Q^P and $v(RzR)>0$ entails z in P, we have $RyRRzR\subset P$, i.e. $v(RyRRzR)>0$, contradiction.
So $v(RyR)\geq 0$, i.e. $y \in \{x \in Q|v(RxR)\geq 0\}$. Suppose now that $(P,Q^P)<(T,Q^T)$.
Take y in Q^T-Q^P and x in Q such that $xRyR\subset Q^P-P$, then directly from the definition of a prime it follows that x is in P. But $x \in P\subset T$, $RyR\subset Q^P$ so $xRyR\subset T\cap Q^P=P$, a contradiction.

Corollary 11. *Let* (P,Q^P) *be a prime in* Q, $R\subset Q^P$, *write* $p=P\cap R$ *for the corresponding prime ideal in* R. *Then* $(P,Q^P)=(Q_{R-p}(R)p,Q_{R-p}(R))$.

Proof: In view of corollary 9. it is easy to verify that $(Q_{R-p}(R)p, Q_{R-p}(R))$ dominates (P,Q^P). By the above proposition the result follows.

Remark. 1) The above corollary also makes it possible to obtain the primes (P,Q^P) from "p-adic" a.p.v.'s:
Let p be a prime ideal in R, and I a fractional ideal, then $I=p^{n_p}S$, S an element of $\mathcal{F}$ and p not occuring as a factor in the unique factorisation of S. The function $v_p:\mathcal{F} \to \mathbb{Z}$; $I \to n_p$ defines an a.p.v. on Q, with $R\subset Q^{P_{v_p}}$. Straightforward computation then shows that $(P_{v_p},Q^{P_{v_p}})=$

$(Q_{R-p}(R)p, Q_{R-p}(R))$.
2) The set of all inequivalent a.p.v. on R, V, satisfies the approximation property:
Let $\{v_1,\ldots,v_t\}$ *be a finite set of a.p.v. in* V, $\{n_1,\ldots,n_t\}$ *a finite subset of* $\mathbb{Z}$ *and* $\{a_1,\ldots,a_n\}$ t *elements of* Q; *then there exists an element* x *in* Q *such that* $v_i(x-a_i)\geq n_i$, $i=1,\ldots,t$ *and* $v(x)\geq 0$ *for all* v *in* $V-\{v_1,\ldots,v_t\}$.
3) This theory of a.p.v. in Q yields an alternative approach for the ideal theory of arithmetical rings (e.g. Asano orders). This follows from the following result:
If V *is a set of inequivalent a.p.v. in* Q, *satisfying the approximation property, then the ring* $\bigcap_{v\in V} Q^v$ *is an arithmetical ring.*
(cf. (4) for more details).

3. Examples.

Let A be a simple ring, σ an automorphism of A, and δ a σ-derivation of A. Consider the skew polynomial ring $R=A[X,\sigma,\delta]$. Using results of G.Cauchon, cf. (1), one obtains:

Theorem 12. *If* R *has non-trivial center* Z(R), *i.e.* $Z(R)\neq K=\{a\in Z(A) \mid \delta(a)=0,\ \sigma(a)=a\}$, *then* R *is an arithmetical ring.*
Proof: cf.(3).

Note that the hypothesis is necessary.
Let $R=A[X,\sigma,\delta]$ be non simple and let the center of R be trivial. Take f a non-constant, monic polynomial of minimal degree, such that Rf is an ideal of R. Then Rf is the unique maximal ideal of R and all non-trivial ideals of R are of the form Rf^m, $m>0$ in $\mathbb{N}$. Since Rf and (0) are both lying over (0) in K (=Z(R)), R is not a Zariski extension of its center.
Independently of the above Theorem it is possible to prove:
If $R=A[X,\sigma,\delta]$ *has a classical left ring of quotients* Q, *then* R *is an Asano left order in* Q. *In case* A *is a simple Artinian ring then it follows that* $A[X,\sigma,\delta]$ *is a Dedekind prime ring.*
So if moreover R has non-trivial center then $A[X,\sigma,\delta]$ is an Asano left order (Dedekind prime ring) which is a global Zariski extension of its center.
The following example shows that Dedekind prime rings need not to be global Zariski extensions of their center.
Take A a simple Artinian ring, let σ be an automorphism of A no power

of which is inner. Then $A[X,\sigma,-]$ is a non-simple Dedekind prime ring with trivial center. By the above it is not a Zariski extension of its center.

References.

(1) G.Cauchon: Les T-anneaux et les anneaux à identités polynomiales Noethériens; Thèse Univ. de Paris XI - Centre d'Orsay 1977.

(2) E.Nauwelaerts: Zariski extensions of rings; Ph.D-thesis Antwerp 1979.

(3) E.Nauwelaerts - F.Van Oystaeyen: Birational Hereditary Noetherian prime rings; Comm. in Alg. 8(4), 309-338 (1980).

(4) J.Van Geel: Primes and Value functions; Ph.D-thesis Antwerp 1980.

(5) J.Van Geel: Primes in Algebras and the Arithmetic of Central Simple Algebras; Comm. in Alg. 8(11), 1015-1053 (1980).

(6) F.Van Oystaeyen: Zariski Central Rings; Comm. in Alg. 6(8), 799-821 (1978).

(7) F.Van Oystaeyen: Prime spectra in non-commutative algebra; Lect. Notes in Math. 444, Springer Verlag Heidelberg-Berlin 1975.